Excel Guide for Finite Mathematics and Applied Calculus

Second Edition

Revathi Narasimhan
Kean University

A technology guide to accompany

Mathematical Applications, 7th Edition
Harshbarger/Reynolds

Calculus: An Applied Approach, 6th Edition
Brief Calculus: An Applied Approach, 6th Edition
Brief Calculus: An Applied Worktext
Larson/Edwards

Applied Calculus, 3rd Edition
Brief Applied Calculus, 3rd Edition
Berresford/Rockett

Calculus with Finite Mathematics
Berresford/Rockett

Finite Mathematics
Berresford/Rockett

HOUGHTON MIFFLIN COMPANY BOSTON NEW YORK

Sponsoring Editor: Lauren Schultz
Assistant Editor: Marika Hoe
Senior Manufacturing Coordinator: Florence Cadran
Senior Marketing Manager: Danielle Potvin

Printed in the U.S.A.

ISBN: 0-618-29360-4

1 2 3 4 5 6 7 8 9 - CRS - 06 05 04 03 02

Preface

The use of spreadsheets in a course designed primarily for business and social science majors can enhance the understanding of the underlying mathematical concepts. In addition, the widespread use of spreadsheets in these fields makes it a natural vehicle for students to experiment with the ideas they have learned. Prior knowledge of Excel is not assumed for use of this supplement. Students are provided enough introductory material to get started, and specialized instructions are given as needed.

Excel Guide to Finite Mathematics and Applied Calculus, 2nd ed., is designed as a technology supplement for the following textbooks:

Mathematical Applications, 7th ed.
Harshbarger-Reynolds

Calculus: An Applied Approach, 6th ed.
Brief Calculus: An Applied Approach, 6th ed
Brief Calculus: An Applied Worktext
Larson-Edwards

Applied Calculus, 3rd ed.
Brief Applied Calculus, 3rd ed.
Berresford-Rockett

Calculus with Finite Mathematics
Berresford-Rockett

Finite Mathematics
Berresford-Rockett

Software Requirements

This manual was written for Excel 2000, although it can be used with Excel 97 with no changes. Older versions of Excel, such as Excel 95 or Excel 5.0 can also be used, but the graphing directions and screens will be somewhat different. Screenshots use Excel 2000. All instructions in the manual use only the native capabilities of Excel. There is no need for external macros.

The only add-in that is required is Solver, which is used for the linear programming and optimization sections. Solver comes with the standard distribution of Excel. To check if Solver is installed, click on the **Tools** menu and see if Solver appears in the full menu. If it does, you are all set. If not, you will have to add it in. To add it in, click on **Tools > Add-ins**. If a box for Solver appears, check it, and click OK. If Solver does not appear in the add-ins list, it must be installed through the setup program for Excel. See your systems administrator for further details.

Correlation Guide for Textbooks

The following tables outline the relationship between the material in this manual and topics covered in the textbooks.

Correlation table for *Mathematical Applications, 7th ed.*, by Harshbarger-Reynolds

Chapter in Excel Guide	Chapter(s) in Text	Comments
1 Getting Started	N/A	Excel material needed for all subsequent sections
2 Graphs of Functions	Chapters 1 and 2	Excel material needed for all subsequent sections
3 Linear and Polynomial Regression	Section 2.5 and various modeling exercises	Needed for modeling exercises appearing in many chapters
4 Finding Zeros with Goal Seek	Section 2.3	Intercepts and break-even problems; solving equations
5 Matrices	Chapter 3	Can also be used for Markov chains
6 Linear Programming using Solver	Chapter 4	
7 Mathematics of Finance	Chapter 6	
8 Probability and Statistics	Chapter 8	
9 Limits and Derivatives	Chapter 9	
10 Graphs of Functions and their Derivatives	Chapter 10: sections 1,2	
11 Optimization in One Variable using Solver	Chapter 10: sections 3,4	
12 Exponential, Log and Trig Functions	Chapter 5	
13 Integration	Section 13.1	
14 Graphs of Functions of Two Variables	Section 14.1	
15 Constrained Optimization	Section 14.5	

Correlation table for *Calculus: An Applied Approach, 6th ed.* by Larson-Edwards

Chapter in Excel Guide	Chapter(s) in Text	Comments
1 Getting Started	N/A	Excel material needed for all subsequent sections
2 Graphs of Functions	Sections 1.1-1.4	Excel material needed for all subsequent sections
3 Linear and Polynomial Regression	Section 7.7, but see comment on right	Needed for modeling exercises appearing in many chapters
4 Finding Zeros with Goal Seek	Section 1.2	Useful for determining intercepts and solving equations
8 Probability and Statistics	Chapter 9	
9 Limits and Derivatives	Sections 1.5,1.6; Chapter 2	
10 Graphs of Functions and their Derivatives	Sections 2.3, 3.1-3.3	
11 Optimization in One Variable using Solver	Sections 3.4,3.5	
12 Exponential, Log and Trig Functions	Chapters 4 and 8	Chapter 8 not in brief versions of text
13 Integration	Sections 5.6 and 6.5	
14 Graphs of Functions of Two Variables	Section 7.3	
15 Constrained Optimization	Section 7.6	
16 Taylor Series	Sections 10.4,10.5	
17 Newton's Method	Section 10.6	

Correlation table for *Brief Calculus: An Applied Approach*, *6th ed.*, and *Brief Calculus: An Applied Worktext*, by Larson-Edwards

Chapter in Excel Guide	Chapter(s) in Text	Comments
1 Getting Started	N/A	Excel material needed for all subsequent sections
2 Graphs of Functions	Sections 1.1-1.4	Excel material needed for all subsequent sections
3 Linear and Polynomial Regression	Section 7.7, but see comment on right	Needed for modeling exercises appearing in many chapters
4 Finding Zeros with Goal Seek	Section 1.2	Useful for determining intercepts and solving equations
9 Limits and Derivatives	Sections 1.5,1.6; Chapter 2	
10 Graphs of Functions and their Derivatives	Sections 2.3, 3.1-3.3	
11 Optimization in One Variable using Solver	Sections 3.4,3.5	
12 Exponential, Log and Trig Functions	Chapter 4	
13 Integration	Sections 5.6 and 6.5	
14 Graphs of Functions of Two Variables	Section 7.3	
15 Constrained Optimization	Section 7.6	

Correlation table for *Applied Calculus, 3rd ed.* by Berresford-Rockett

Chapter in Excel Guide	Chapter in Text	Comments
1 Getting Started	N/A	Excel material needed for all subsequent sections
2 Graphs of Functions	Chapter 1	Excel material needed for all subsequent sections
3 Linear and Polynomial Regression	Section 7.4, but see comment on right	Needed for modeling exercises appearing in many chapters
4 Finding Zeros with Goal Seek	Section 1.3, 1.4	Break-even problems; solving equations
8 Probability and Statistics	Sections 11.1,11.4	
9 Limits and Derivatives	Sections 2.1,2.2	
10 Graphs of Functions and their Derivatives	Sections 3.1,3.2	
11 Optimization in One Variable using Solver	Sections 3.3-3.5	
12 Exponential, Log and Trig Functions	Chapters 4 and 8	
13 Integration	Sections 5.3, 5.4, and 6.4	
14 Graphs of Functions of Two Variables	Section 7.1	
15 Constrained Optimization	Section 7.5	
16 Taylor Series	Sections 10.2,10.3	
17 Newton's Method	Section 10.4	

Correlation table for *Brief Applied Calculus, 3rd ed.*, by Berresford-Rockett

Chapter in Excel Guide	Chapter in Text	Comments
1 Getting Started	N/A	Excel material needed for all subsequent sections
2 Graphs of Functions	Chapter 1	Excel material needed for all subsequent sections
3 Linear and Polynomial Regression	Section 7.4, but see comment on right	Needed for modeling exercises appearing in many chapters
4 Finding Zeros with Goal Seek	Section 1.3, 1.4	Break-even problems; solving equations
9 Limits and Derivatives	Sections 2.1,2.2	
10 Graphs of Functions and their Derivatives	Sections 3.1,3.2	
11 Optimization in One Variable using Solver	Sections 3.3-3.5	
12 Exponential, Log and Trig Functions	Chapters 4 and 8	
13 Integration	Sections 5.3, 5.4, and 6.4	
14 Graphs of Functions of Two Variables	Section 7.1	
15 Constrained Optimization	Section 7.5	

Correlation table for *Calculus with Finite Mathematics*, by Berresford-Rockett

Chapter in Excel Guide	Chapter(s) in Text	Comments
1 Getting Started	N/A	Excel material needed for all subsequent sections
2 Graphs of Functions	Chapter 0	Excel material needed for all subsequent sections
3 Linear and Polynomial Regression	Section 11.4, but see comment at right	Needed for modeling exercises appearing in many chapters
4 Finding Zeros with Goal Seek	Section 0.3, 0.4	Intercepts and break-even problems; solving equations
5 Matrices	Chapter 2	Can also be used for Markov chains
6 Linear Programming using Solver	Chapter 3	
7 Mathematics of Finance	Chapter 1	
8 Probability and Statistics	Section 4.4, Chapter 5	
9 Limits and Derivatives	Sections 6.1,6.2	
10 Graphs of Functions and their Derivatives	Sections 7.1,7.2	
11 Optimization in One Variable using Solver	Sections 7.3-7.5	
12 Exponential, Log and Trig Functions	Chapter 8	
13 Integration	Sections 9.3, 9.4, and 10.4	
14 Graphs of Functions of Two Variables	Section 11.1	
15 Constrained Optimization	Section 11.5	

Correlation table for *Finite Mathematics*, by Berresford-Rockett

Chapter in Excel Guide	Chapter(s) in Text	Comments
1 Getting Started	N/A	Excel material needed for all subsequent sections
2 Graphs of Functions	Chapter 1	Excel material needed for all subsequent sections
3 Linear and Polynomial Regression	See comment at right	Needed for modeling exercises appearing in many chapters
4 Finding Zeros with Goal Seek	Section 1.3, 1.4	Intercepts and break-even problems; solving equations
5 Matrices	Chapter 3	Can also be used for Markov chains
6 Linear Programming using Solver	Chapter 4	
7 Mathematics of Finance	Chapter 2	
8 Probability and Statistics	Section 5.6, Chapter 6	

Contents

1 Getting Started With Excel ... 1

Data and Cell References ... 2
Formatting Cells ... 2
Formulas ... 3
Copying and Pasting ... 4
File Operations ... 6
Tables in Excel ... 6
Some Common Errors ... 9

2 Graphs of Functions ... 10

Graphing a Single Function ... 10
Graphing More than One Function ... 13
Graphing Options ... 14
Graphing Discontinuous Functions ... 15
Plotting Functions using Cases ... 16

3 Linear and Polynomial Regression ... 18

Linear Regression using Chart Wizard ... 18
Linear Regression using Excel Functions ... 21
Comparison of Predicted Data with Actual Data ... 22
Forecasting using Linear Regression ... 22
Polynomial Regression ... 23

4 Finding Zeros of a Function with Goal Seek ... 27

Finding the X-intercept of a Line ... 27
Finding Zeros of a Quadratic Function ... 28
Break-even Problems using Goal Seek ... 31

5 Matrices ... 33

Adding and Multiplying Matrices ... 33
Multiplication of Matrices ... 34
Inverse of a Matrix ... 36
Solving Systems of Equations using Inverses ... 37
Leontief Input-Output Model ... 40

6 Linear Programming using Solver ... 41

Maximization Problem using Solver ... 41
Minimization Problem using Solver ... 46

7 Mathematics of Finance ... 50

Simple Interest ... 50
Compounded Interest Using Tables ... 51
Compounded Interest Using Excel Functions ... 53
Future Value of Ordinary Annuities and Annuities Due ... 54
Calculating Payment for Annuities and Sinking Funds ... 56
Present Value of Annuities ... 58
Loans and Amortization ... 59

8 Probability and Statistics ... 63

Binomial Probability ... 63
Descriptive Statistics ... 65
Normal Probabilities ... 70

9 Limits and Derivatives 73
Numerical Investigation of Limits 73
Numerical Investigation of Derivatives 75
Derivatives and Rates of Change 76
10 Graphs of Functions and their Derivatives 78
A Function and its First Derivative 78
A Function and its Second Derivative 80
11 Optimization in One Variable Using Solver 82
12 Exponential, Logarithmic and Trigonometric Functions 87
Graphs of Exponential, Log and Trig Functions 87
Solving Equations with Goal Seek 90
13 Integration 92
Approximating Area using Rectangles - Left Endpoint 92
Approximating Area using Rectangles - Midpoint 93
Numerical Integration - Trapezoidal Rule 95
Numerical Integration - Simpson's Rule 97
14 Graphs of Functions of Two Variables 100
15 Constrained Optimization and Lagrange Multipliers 104
16 Taylor Series 109
17 Newton's Method 112
Finding a Zero of a Function 112
Finding Intersection of Two Functions and Stopping Criteria 114

Chapter 1

Getting Started With Excel

This chapter will familiarize you with various basic features of Excel. Specific features which you need to solve a problem will be introduced as the need arises. When working with the examples given, you should be at a computer with an open, blank Excel workbook.

Start up Excel using instructions in your local installation of the software. You will see the following screen. Familiarize yourself with the various components of the spreadsheet.

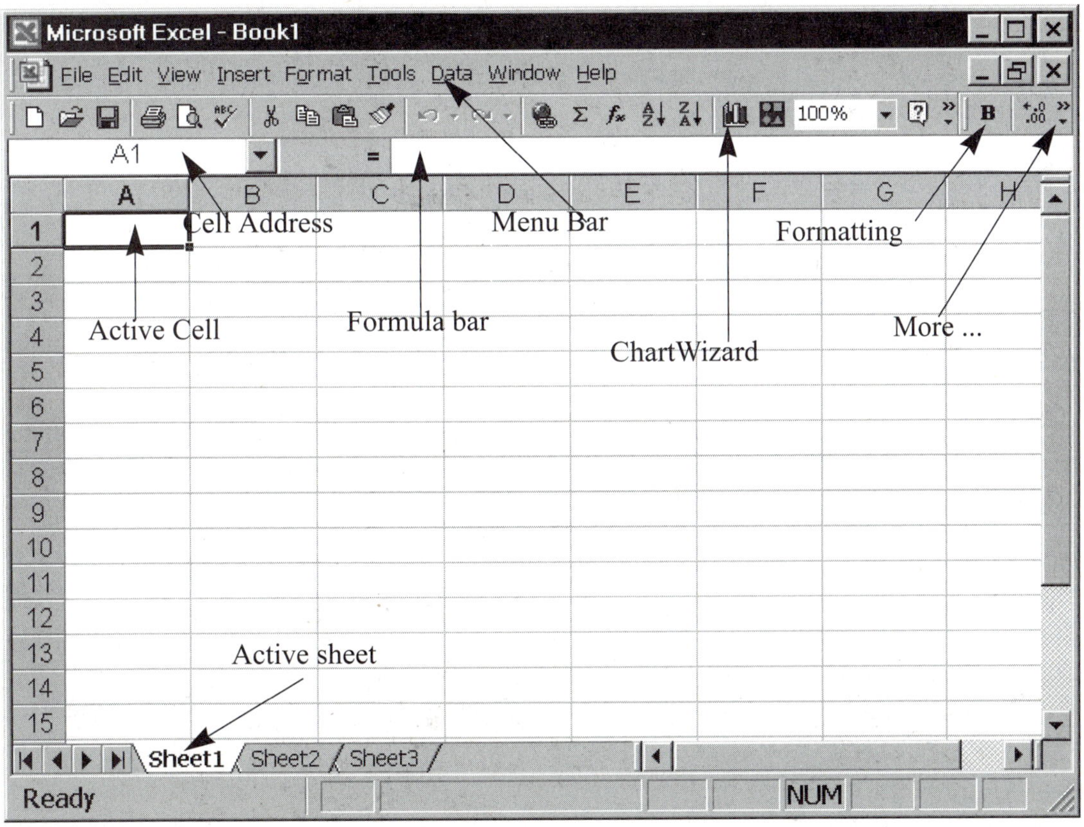

The screen with a grid you are looking at is called a *worksheet*. You can click on the tabs below to go to other worksheets. These worksheets are part of a *workbook* with the file name `book1.xls`, but you can rename it to any file name when you save your file.

If your screen does not exactly resemble the above, don't be concerned. To display all the features shown on this sample worksheet, go to **View > Toolbars** and check the options for standard and formatting toolbars.

Data and Cell References

All information in a spreadsheet is entered through data in cells. Each cell has a unique reference given by its column letter and row number. You will notice that the cell reference box above the column headings says **`A1`**. The reference of the cell can easily be figured out by locating the column and row where it belongs.

To move from one cell to another, you can use the arrow keys or select a cell with a mouse click. You can also type <CTRL>g to go to a specific cell reference.

You can work with a range of cells. To select a range, click into the beginning of the range of cells. Hold down the mouse and drag to the end of the range. Release the mouse button. The reference for a range of cells is given by **`beginning_cell_reference:end_cell_reference`**

Check it out

- Select the range of cells **`h42:j48`**
- Select the range of cells **`b8:d40`**

In the examples given below, a spreadsheet fragment with illustrative cell reference(s) will often appear. These are given to make the examples easier to follow. You can of course use any groups of cells you desire to work the examples, as long you change the cell references to reflect your setup.

Formatting Cells

You can type either text or numbers in a cell. Enter some data by first selecting a cell and typing some text or numbers into it. You can use the back arrow to correct the entry. Press <ENTER>. You may then format the cell content as follows:

1. First select the cell in which some data is entered.
2. Choose the style and size of the font by clicking on the font list.
3. Click on the Bold, Italic or Underline option if you wish to format in one of those styles.
4. Click on the left, center, or right justification for text in a cell.
5. If you have entered a number, you may increase or decrease the number of decimal spaces displayed.

Check it out

- Type in some text in a cell and test out the various formatting capabilities.

Correcting Cell Entries

Once you have entered some data in a cell, and pressed <ENTER>, you may later want to edit it. To do this, press the F2 key. You will see the cursor in the cell. Edit by using the backspace key or by using the mouse cursor. Press <ENTER> to accept the new content.

To delete the contents of the cell, select the cell and press the <DELETE> key. If you want to clear the formatting options from a cell, go to **Edit > Clear > Format**.

Adjusting Cell Width

When you type in text, you may sometimes exceed the width of the cell. To widen a cell, move the mouse along the column you wish to widen to the row with the heading labels at the top of the worksheet. You will see a symbol looking like <-||-> . Holding down the left mouse button, you can now widen the column.

Wrapping Text

For aesthetic reasons, you may not want text in a cell to be too wide. In this case, you must wrap the text within the width of a cell. After selecting the cell, go to **Format > Cells**. Click on the Alignment tab and check the Wrap Text box.

Inserting Rows or Columns

Go to the cell where you want to insert a row or column. Right click the mouse button and choose the Insert option. Click on the appropriate checkbox for inserting rows or columns.

Formulas

Once you have entered data into cells, you will want to perform some operations with them. Basic arithmetic operators are:

Operation	Symbol
Addition	+
Multiplication	*
Division	/
Subtraction	-
Exponentiation	^

The usual order of operations holds. Using the above operators, you can write formulas which manipulate the data you have entered in cells.

Example 1 Let $x = 3$. Compute $f(x) = x^3 - 4x$.

Solution We need to store the x value in a cell. We also need to store the $x^3 - 4x$ result in another cell. Hence, we can make a simple table as follows. Note that you can enter text into a cell as well. Using a spreadsheet makes it easy to annotate your work.

	D	E
1	x	f(x)
2	3	=d2^3-4*d2

Now, the value of x is contained in the cell **D2**. The value for f(x) is computed by the formula using the cell reference **D2** in place of x. So, the formula for f(x) using cell references is **=d2^3-4*d2** (Note: **D2** is the same as **d2**)

To enter this in the spreadsheet:

1. Select the cell **E2**
2. Type the formula **=d2^3-4*d2** in this cell
3. Press <ENTER>

A formula always begins with an = sign. There should be no space before the = sign and there should be no space between the = sign and the rest of the formula.

Now, change the value of x in **D2**. What happens to the value in **E2**?

Check it out

- Change f(x) to $f(x) = 2x^2 + 1$. Enter this formula in **E2** using cell references.
- Be careful when entering formulas. Let the value in **a1** equal some number not equal to 1. What is the output of f(x)=1/(x-1) when incorrectly using the formula **=1/a1-1**? Compare with the correct formula **=1/(a1-1)**

When you look at a worksheet, you cannot see which cells have formulas and which have numbers. If you want to see all the formulas in the spreadsheet in their respective cells, click **Tools > Options**. Click on the View tab and check the Formulas box. Close the dialog box, and now your spreadsheet will display all formulas. To go back to the original view, simply unclick the formula box in the View tab.

Check it out

- Display the formula view for the worksheet above.

Copying and Pasting

Now suppose you want to compute *f(x)* in Example 1 for *x* =1,2,3,4,5. You also want to display all these values simultaneously by creating a table. Instead of typing the formula over and over again, we can copy and paste. This is illustrated in the next example.

Example 2 Compute *f(x)* for x=1,2,3,4,5 and display the results in a table.

Solution Make columns for x and f(x). Enter the x values that you are interested in:

	D	E
1	x	f(x)
2	1	
3	2	
4	3	

	D	E
5	4	
6	5	

In the cell **E2**, enter the formula for $f(x)=x^3-4x$. This gives the following:

	D	E
1	x	f(x)
2	1	**=d2^3-4*d2**
3	2	
4	3	
5	4	
6	5	

Press <ENTER> after entering the formula, and you will see the value of $f(1)=-3$ in the cell **E2**.

Since we want to compute the values of f(x) for the other values of x as well, we can copy the formula by following the steps below.

Copying a formula down a column

1 Select the **E2** cell in the above table. Press <CTRL>c to copy.
2 Select the rest of the f(x) column, cells **E3:E6**. Press <CTRL>v to paste.

Look in the formula bar for the entries **E3:E6** and note that the cell references automatically change to the x-value directly to the left of the y-value.

Autofill option

Another way to copy the formula down the column is to move your mouse to the lower right hand corner of the cell **E2** until you see a small + sign. Then, holding down the left mouse button, drag your mouse down the column to **E6**. The formulas will be automatically changed to reflect the new f(x)-values.

Your table will look like the following, regardless of the method you use to copy the formula:

	D	E
1	x	f(x)
2	1	-3
3	2	0
4	3	15
5	4	48
6	5	105

Check it out

- Change f(x) to $f(x)=-x^2+4$. Remember to recopy the new formula down the column.Somtimes, Excel does not recognize the (-) sign in front of an expression. To be on the safe side, enter the formula as =(-1)*d1^2+4.

File Operations

Now that you have entered various items in your workbook, you will want to save and/or print the file. The following table summarizes how to perform various operations with your Excel file.

Operation	How to perform
Open new file	**Menu > File > New**
Open old file	**Menu > File > Open**; then follow dialog box
Saving new file	**Menu > File > Save As**; then follow dialog box
Saving to current file	**Menu > File > Save** or <CTRL>s
Printing file	**Menu > File > Print** or <CTRL>p
Page setup	**Menu > File > Page Setup**

Print preview and formatting your worksheet

Within the page setup option outlined above, you can set headers, footers, margins, and orientation of the page (portrait or landscape). You can then use **File > Print Preview** to preview your final output. Although it is preferable to have the grid lines visible on the computer, you should normally not print out the grid lines. The default option in Excel 2000 is to suppress the printing of gridlines.

You may want to outline your tables with borders. The border formatting icon on the formatting bar will show you various options.

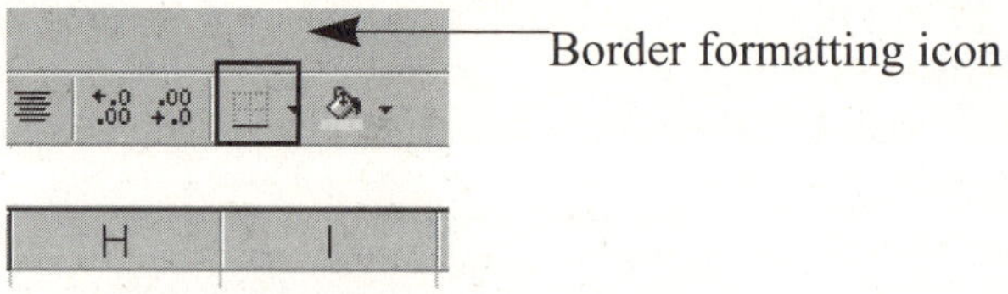

Tables in Excel

In order to use the graphing features of Excel, you will first need to generate tables of x and y values. In this section, you will learn how to easily generate equally spaced entries for use as x-values.

Example 1 Generate a table of values from -2 to 3 in increments of 0.5.

Remark We could of course do this manually, but that would be laborious. Excel can automatically generate this table by using the Fill feature.

Solution

Steps to create a table of x-values

1 Type a heading label **x** in cell **A1**.
2 Type in the first value of -2 in the cell **A2**.
3 In cell **A3**, type in the next value of -1.5, since our increments are in steps of 0.5. Now that you have entered a starting value and a value with the increment, Excel can generate the rest of the table.

	A
1	x
2	-2
3	-1.5

4 Select the cells **a2:a3** . Move mouse to lower right corner until you see a plus sign. Your screen should resemble the following figure:

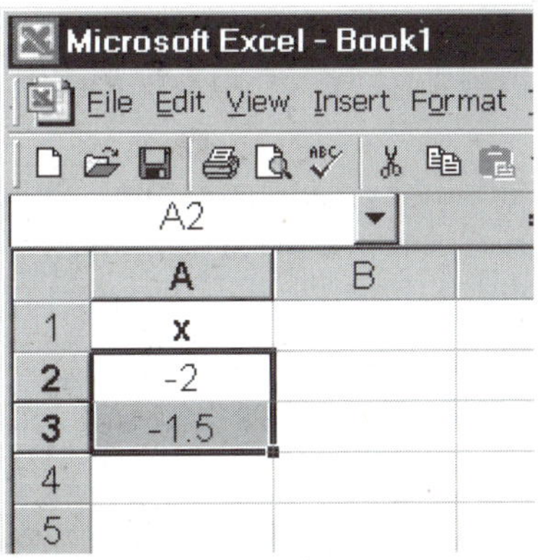

5 Drag the mouse all the way down the column to **A12**. You should now see a filled column of values from -2 to 3 in increments of 0.5, like the one below.

	A
1	x
2	-2
3	-1.5
4	-1
5	-0.5
6	0
7	0.5
8	1.0

	A
9	1.5
10	2
11	2.5
12	3

Example 2 Suppose we want to generate x and y values in a table. For example, find $f(x) = 3x - 2$ for the x-values given in the table above.

Solution Follow the steps outlined below.

Steps for creating table with x and y values

1 Make a table with x and f(x) column headings.
2 Fill the x-column as directed in Example 1.
3 Next, we need to fill in values for f(x).

  a The first y-value will have the formula **=3*a2-2**. Type it into the cell **B2**.

	A	B
1	x	f(x)
2	-2	**=3*a2-2**

  b We next fill the entire f(x) column by simply copying and pasting the formula in the **B2** cell: Select the **B2** cell in the above table. Press <CTRL>c to copy.
  c Select the rest of the f(x) column. Press <CTRL>v to paste. Note that the cell references automatically change to the x-value directly to the left of the y-value.

4 Alternatively, we could have used the Autofill method outlined in the Copying and Pasting section.
5 Your table should resemble the one below

	A	B
1	x	f(x)
2	-2	-8
3	-1.5	-6.5
4	-1	-5
5	-0.5	-3.5
6	0	-2
7	0.5	-0.5
8	1.0	1.0
9	1.5	2.5
10	2	4
11	2.5	5.5

	A	B
12	3	7

Check it out

- Change f(x) to $f(x)=-x^2+4$. Remember to recopy the new formula down the column. Sometimes, Excel does not recognize the (-) sign in front of an expression. To be on the safe side, enter the formula as =(-1)*a1^2+4.
- Create a table of x and y values for f(x)=2x-4 for values of x between -2 and 3 in increments of 1.

Some Common Errors

Grayed out option boxes

This happens when you try to do something with a cell, but are still working with that cell. Click out of the cell and click back in and now select the option.

#REF, #####, #DIV/0 and other error messages

#REF usually indicates an erroneous cell reference. Check your formulas in formula view if necessary.

means that the number did not fit in the cell. Simply widen the cell to suitable width.

#DIV/0 means you're dividing by zero. Check your formulas and their references.

#NAME? usually indicates an invalid name of a function.

Described menu option does not appear in pull-down menu

This is not really an error, but rather a "feature" in Excel 2000. If you do not see a menu option listed in the pull-down menu, you should double click on the menu title or click on the double arrows pointing down - it will then list the complete set of options.

Chapter 2

Graphs of Functions

Graphing a Single Function

To graph functions in Excel, you must first create a table of data with the information about the x and y values. You then use Chart Wizard to create the plot. The following example will take you through the process step by step.

Example 1 Graph the function $f(x)=2x^2+x$ on the interval [-2,2].

Solution Follow the steps outlined below.

Creating the graph of a function

1 First create a table of x and y values as explained in the Tables section of Chapter 1. The y-values are given by the formula for $f(x)$. The formulas for the first two y-values are given as an illustration.

	A	B
1	x	f(x)
2	-2	`=2*a2^2+a2`
3	-1.5	`=2*a3^2+a3`
4	-1	1
5	-0.5	0
6	0	0
7	0.5	1
8	1	3
9	1.5	6
10	2	10

2 Select the range of x and y values which you wish to plot. For this example, it is the range **`a1:b10.`** We select the column headings as well as the numbers.

3 Click on the ChartWizard icon, which is on the menu bar with a bar graph icon. You will see the following dialog box.

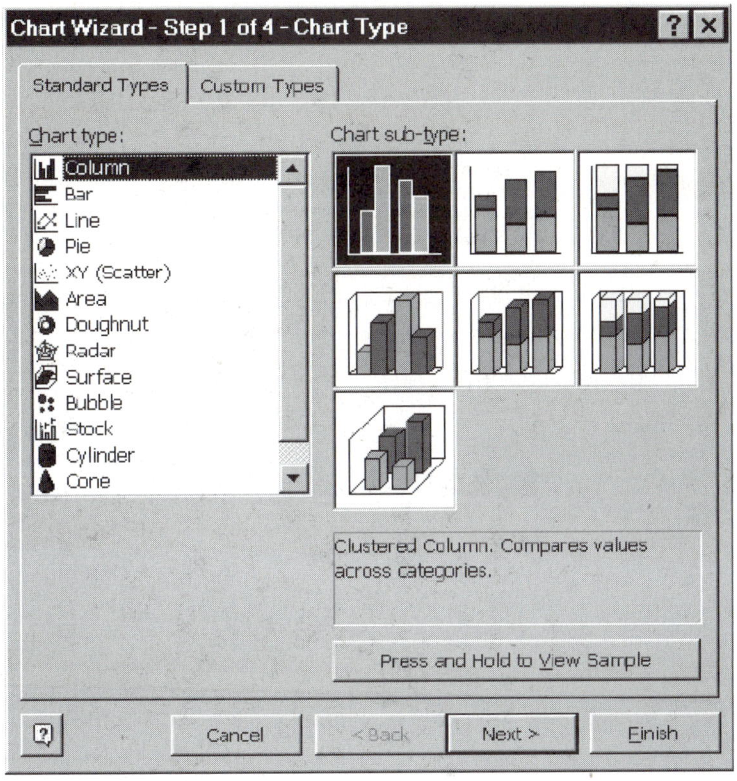

4 Select the X-Y (Scatter) chart type with the smoothed line with no markers subtype. The dialog box will then look as follows:

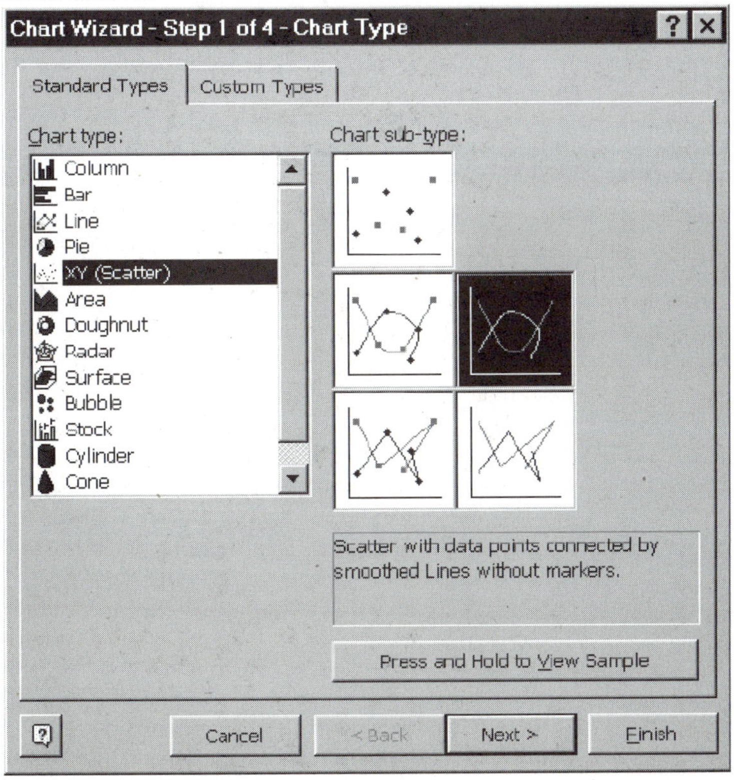

5 Click the Next button. You will see the following dialog box. Examine the boxes to see the correct information is entered.

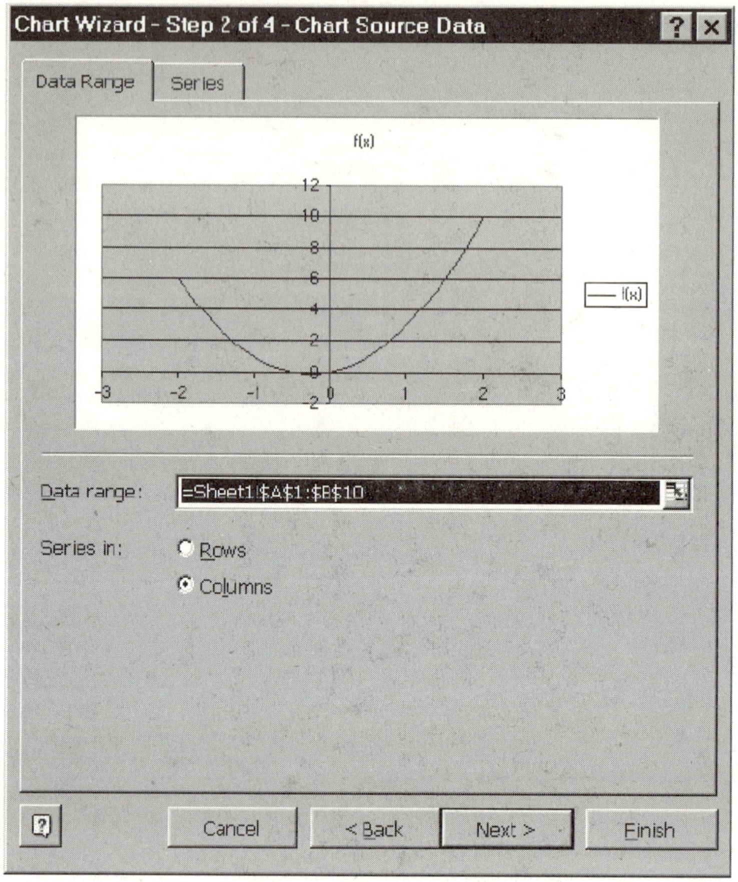

6 Click the Next button. You will see a dialog box for Step 3. Enter labels for the x-axis, y-axis as well as a title for the graph. Click Next.

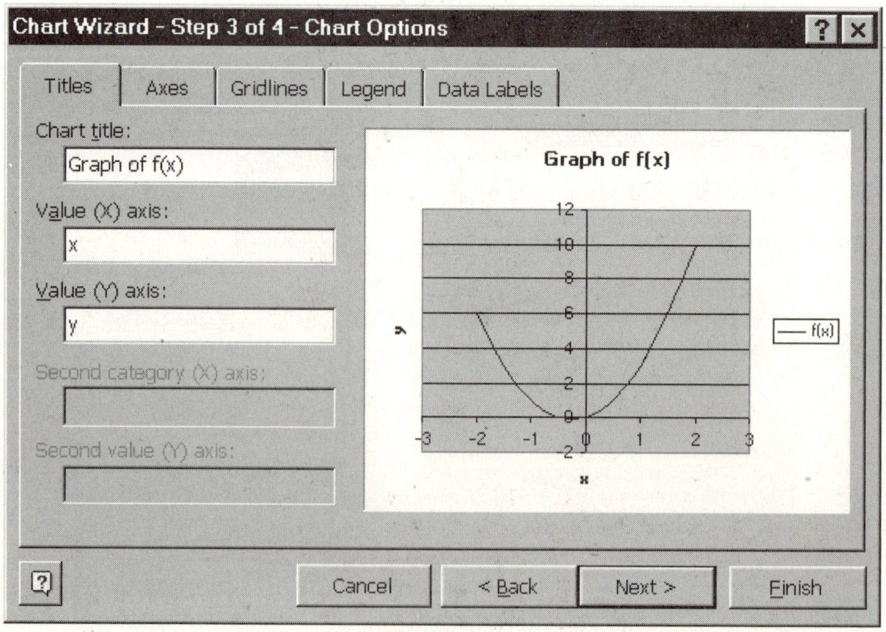

7 You are now at Step 4 of the ChartWizard, the final step. In this step, you will be asked where to place the graph. By default, the option to place the chart within the current worksheet will be checked. Just click Finish and you will see a graph like the following.

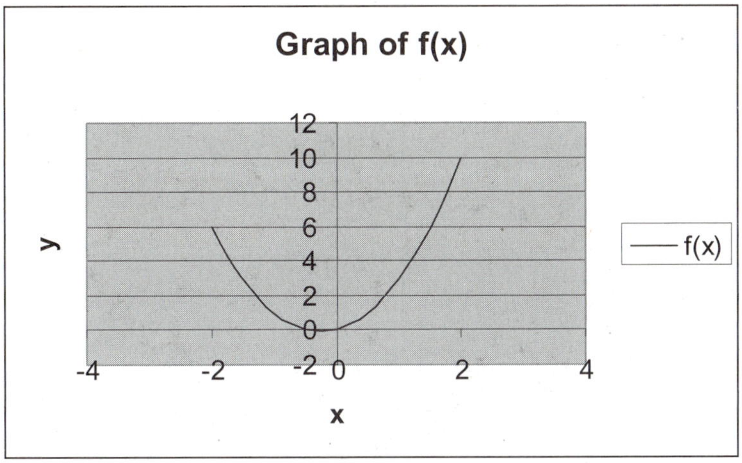

Check it out

- Graph the function $f(x)=2x^3$ on the interval [-2,1].

Graphing More than One Function

To graph more than one function on the same plot with the same range of x-values, simply create a table with multiple column headings, with one heading for each function. The next example illustrates this.

Example 2 Graph $f(x)=x^2$ and $g(x)=x^3$ on the interval [-2,2].

Solution Follow the steps outlined below to graph more than one function.

Steps to graph more than one function on the same plot

1 Create the following table with x-spacing of 0.5. Create the values for f(x) and g(x) with formulas. The formulas are given in the first row as an illustration. Note that f(x) and g(x) each have a separate column.

	A	B	C
1	x	f(x)	g(x)
2	-2	=a2^2	=a2^3
3	-1.5	2.25	-3.375
4	-1	1	-1
5	-0.5	0.25	-0.125
6	0	0	0
7	0.5	0.25	0.125

	A	B	C
8	1	1	1
9	1.5	2.25	3.375
10	2	4	8

2 Select the range of cells `A1:C10`.
3 Click on the ChartWizard icon and follow steps 3-7 in the previous section. You will get the following graph.

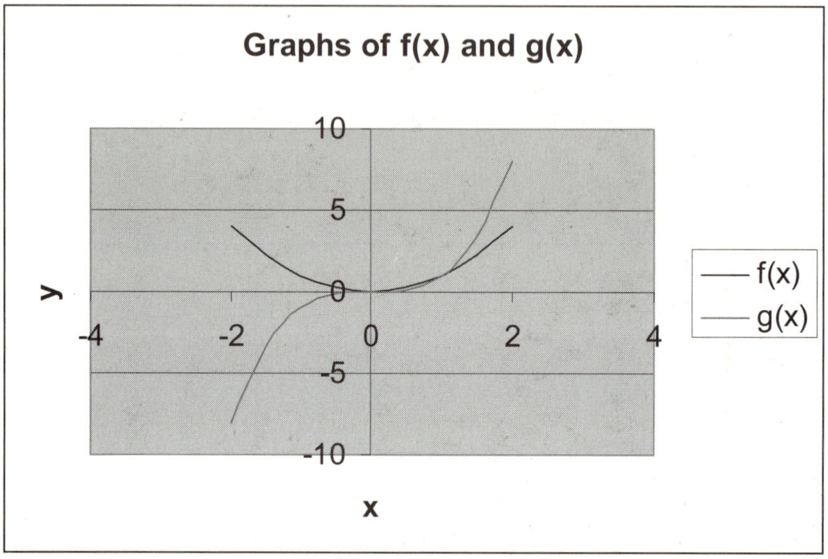

Graphing Options

Excel has many options to adjust the way your plot looks. Once you have placed the chart in the worksheet, you can right click into the chart to list a variety of options. Here, we will briefly outline some major options.

Changing scale

You may want to adjust the scale on the x and/or y axes.

Changing scale on chart in previous example

1 Double click the x-axis. You will see a dialog box.
2 Click on the Scale tab.

3 Change the Minimum and Maximum values and uncheck the corresponding Auto boxes, as shown below:

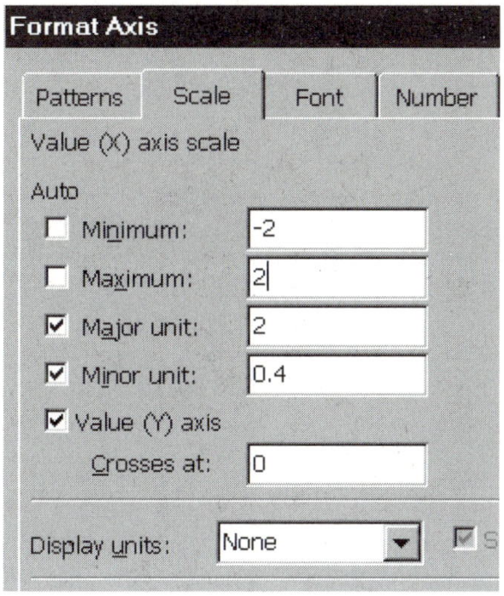

4 Click OK and the x-axis scale will be adjusted on your chart
5 You can do a similar scaling on the y-axis.

Changing marker styles

To change the line color and or marker styles in a plot, double click into the curve you want to change. You will get a dialog box labeled Format Data Series. Click on the Patterns tab and choose the options you desire.

Deleting background shading

By default, Excel provides background shading in all the charts. This can at many times obscure the graphs you generate. To get rid of the shading, double click into the plot area and check "None" for format of plot area. Many of the graphs generated in this manual have used this option to get rid of the background shading.

Graphing Discontinuous Functions

When graphing functions in Excel, all the values listed in the table will be connected together by a curve. If a graph is discontinuous at some point at some x-value, you must leave the corresponding f(x) value blank. Then, Excel will not connect the values.

Example Graph the function $f(x)=1/(1-x)$.

Solution Note that this function is discontinuous at $x=1$.

Steps to plot a discontinuous function

1 Generate a table for x from -1 to 1 and from x=1 to 3, following the directions for generating tables in the previous sections. You will want to have more points near x=1.
2 Generate the f(x) values and remember to leave a blank cell for the f(x) value for x=1.

3 Your table will resemble the following:

x	f(x)
-1	-0.5
-0.5	-0.66667
0	-1
0.5	-2
0.75	-4
0.9	-10
1	
1.1	10
1.25	4
1.5	2
2	1
2.5	0.666667
3	0.5

4 Note that there is no f(x) value for x=1.

5 Select the table and plot using ChartWizard. You will get a plot similar to the following:

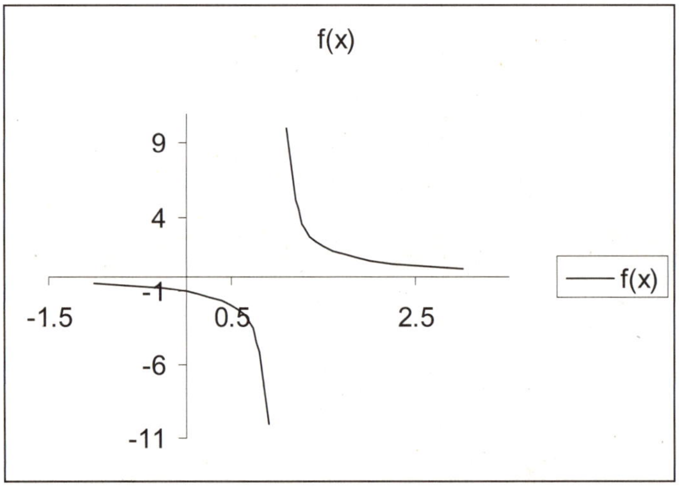

Check it out

- Plot the function $f(x)=1/(x-2)^2$

Plotting Functions using Cases

Sometimes, functions will have different definitions depending on the domain. To generate the table of values for such functions, simply enter the appropriate formula for f(x) for the corresponding x-values. You should be careful not to blindly copy a single formula down the entire column for these types of functions.

Suggested Exercises

Text	Section(s)	Exercises
Harshbarger-Reynolds *Mathematical Applications,* 7th ed.	1.4 2.1-2.4	#1-34, 39-52 Graphing utility problems
Larson-Edwards *Calculus: An Applied Approach,* 6th ed. (and brief version)	1.1 1.2	#33,34 #7-12, 47-54
Berresford-Rockett *Applied Calculus,* 3rd ed. (and brief version)	1.3 1.4	#31-34, 63,64 #53,54, 84-86, 95
Berresford-Rockett *Calculus with Finite Mathematics*	0.3 0.4	#31-34, 63,64 #53,54, 84, 91,92
Berresford-Rockett *Finite Mathematics*	1.3 1.4	#31-34, 63,64 #53,54

Chapter 3

Linear and Polynomial Regression

Many applications of mathematics involve data which must be fitted with a function that best expresses the relationships between the variables in the data set. This chapter will show you how to use Excel to find best fit lines and polynomials.

Linear Regression using Chart Wizard

Example The following spreadsheet table gives the percentage of people 18 and over who used the internet at home in 1997. The breakdown is based on their annual income. (Source: U.S. Census Bureau.)

	A	B
1	Annual Income	Percentage
2	12,500	25.4
3	17,500	28.8
4	22,500	27.9
5	27,500	29.5
6	40,000	31
7	62,500	36.5
8	75,000	43.7

Find the relationship between the percentage of people who use the internet at home and their annual income. That is, plot this set of data and find the line of best fit.

Solution

Part A: Scatter plot

1 Make a scatterplot of the data by selecting the cells containing the data (including the headings) and then click on ChartWizard. Choose the X-Y Scatter with markers.

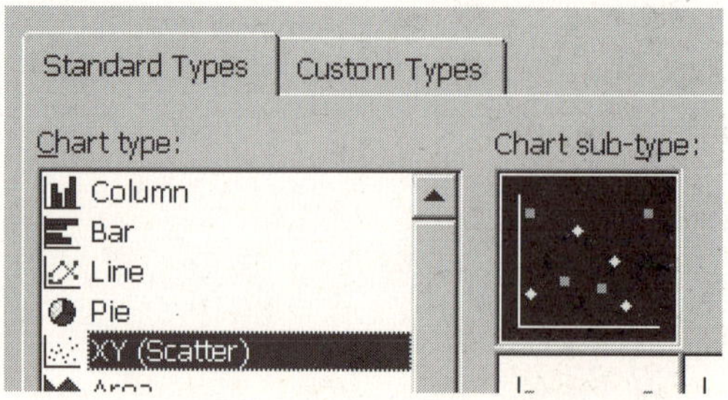

2 Click Next. You will now see a preview of your plot. Note that income and percentage of people who use the internet seem to display a linear relationship. Click Next.
3 You will now be at Step 3 of ChartWizard. Enter suitable text for title and x and y axes name. Change any options, if you wish. Otherwise, simply click Next. You are now at Step 4 of Chart Wizard. Click on the option to place the chart in the current worksheet to get a picture similar to the one below.

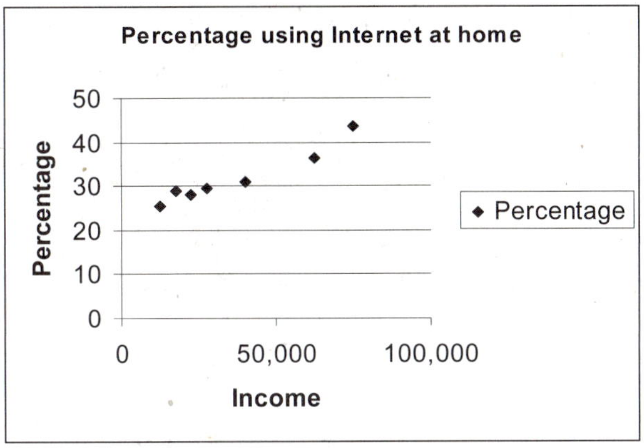

Part B: Adding Trendline

You are now ready to add the line of best fit to this chart using the following steps.

1 Single click into the chart in your workbook.
2 From the Chart menu, choose Add Trendline. You will then see a dialog box like the one below. Click on the Linear option for Trend/Regression type.

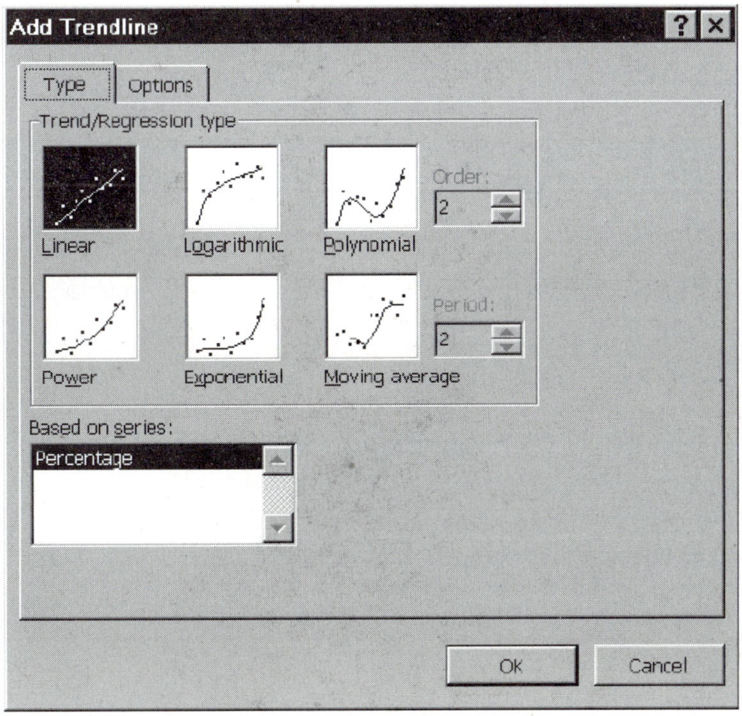

3 Click into the Options tab. Check the box for displaying equation on chart to see a dialog box like the one below.

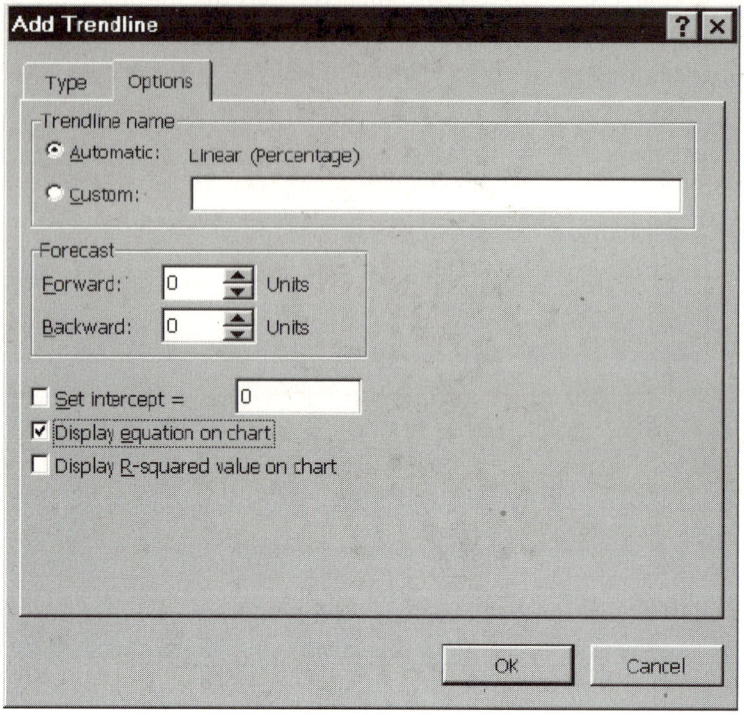

4 Click OK and you will see the line of best fit along with its equation as follows.

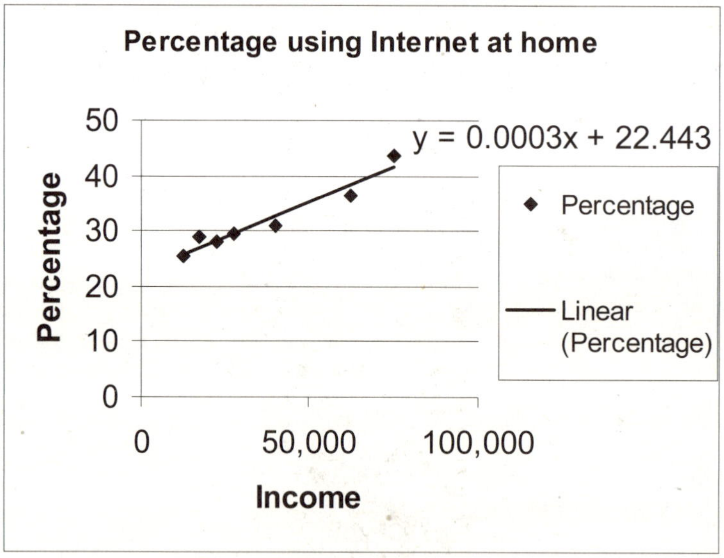

From the inserted text in the chart, we see the equation of the line is y = 0.0003x + 22.443. You can move the equation in the chart into a more viewable position by clicking into it.

Linear Regression using Excel Functions

Using the ChartWizard allows you to visualize data set and the line of best fit. However, you may need to use the equation that is output on the graph elsewhere. Hence, it is useful to be familiar with the built-in Excel functions **`slope`** and **`intercept`**, which give you the slope and y-intercept of the best fit line. You can then use this information in other places in the worksheet.

Example For the data in the example above, use Excel's built-in functions slope and intercept to find the slope and y-intercept of the best fit line.

Solution The data table is reproduced below for easy reference.

	A	B
1	Annual Income	Percentage
2	12,500	25.4
3	17,500	28.8
4	22,500	27.9
5	27,500	29.5
6	40,000	31
7	62,500	36.5
8	75,000	43.7

We calculate slope and the intercept for the line of best fit by typing their formulas in the cells **`E2`** and **`F2`**. The syntax is as follows:

slope(range of y-values,range of x-values)

For this example, the formula would read **`=slope(b2:b8,a2:a8)`**

intercept(range of y-values,range of x-values)

For this example, the formula would read **`=intercept(b2:b8,a2:a8)`**

Typing the formulas for slope and intercept into cells **`E2`** and **`F2`**, respectively, we get the following output:

	E	F
1	slope	intercept
2	0.000255	22.44

Hence, the equation for the line of best fit is y=0.000255x + 22.44. Using the slope function gave the value of the slope to more decimal places than the one given in the chart by adding the trendline.

Comparison of Predicted Data with Actual Data

To see how well the linear function approximated the given data, we next compare the y-values from the data with those predicted by the best fit line. You are using the equation y=mx+b, where m is the slope (in cell **E2**) and b is the y-intercept (in cell **F2**).

Steps for comparison of data

1. Type the heading "Predicted y-value" in the cell **c1** .
2. In cell **c2**, type the formula **=$E$2*A2 + $F$2** . Here, **$E$2** is the slope reference and **$F$2** is the y-intercept reference. **A2** contains a value of x. NOTE: We use the *absolute references* **$E$2** and **$F$2** instead of **E2** and **F2** because we do not want the references to the slope and intercept to change when the formula is copied down the column.
3. Copy the formula in **c2** to **c3:c8**. Your table should look like the following.

Annual Income($)	Percentage	Predicted Percentage	slope	intercept
12,500	25.4	25.63218997	0.000255	22.44288
17,500	28.8	26.90791557		
22,500	27.9	28.18364116		
27,500	29.5	29.45936675		
40,000	31	32.64868074		
62,500	36.5	38.38944591		
75,000	43.7	41.57875989		

We observe that the predicted y-values are fairly close to most of the y-values in the original data set.

Forecasting using Linear Regression

We may also use the linear equation generated by the linear regression method to forecast internet use from home for someone whose annual income is $85,000. We assume that you are using the same spreadsheet from the internet example.

Steps to forecast

1. In cell **A10**, type **85000**
2. Select cell **C8** and copy the formula in **C8** using <CTRL>c
3. Select cell **C10**.
4. Paste the formula in **c10** using <CTRL>v . You will get a percentage of 44.13 . This means that roughly 44% of people with an income of $85,000 use the internet from home, if we use a linear equation model.

Check it out

- Forecast the percentage of people who use the internet from home with an annual income of $95,000

Polynomial Regression

For many data which occur in applications, a linear fit may not be appropriate. You may need to use a best quadratic fit or cubic fit. The next example shows how to fit a polynomial through a set of data of data points.

Example (Data from Example 4, Sec. 7.7, Larson-Edwards)

The number of music CD's, in millions, sold from 1984 through 1994 are listed in the following table. Find the best fit quadratic for the data and use the result to estimate the number of CD's sold in 1998.

Year	1984	1985	1986	1987	1988	1989	1990	1991	1992	1993	1994
y	5.8	22.6	53.0	102.1	149.7	207.2	286.5	333.3	407.5	495.4	662.1

Solution Follow the steps outlined below to create a best fit quadratic.

Part A: Creating a scatterplot

1 Make a table of values of x and y. In order to avoid large numbers, let x=0 correspond to the year 1984. Entering the data in Excel, we get

	A	B	C	D	E	F	G	H	I	J	K	L
1	x	0	1	2	3	4	5	6	7	8	9	10
2	y	5.8	22.6	53.0	102.1	149.7	207.2	286.5	333.3	407.5	495.4	662.1

2 Follow steps 1-3 in the creating scatterplot section of the linear regression section in this chapter. Note that the data are in rows for this example.

3 You will see a scatterplot as follows

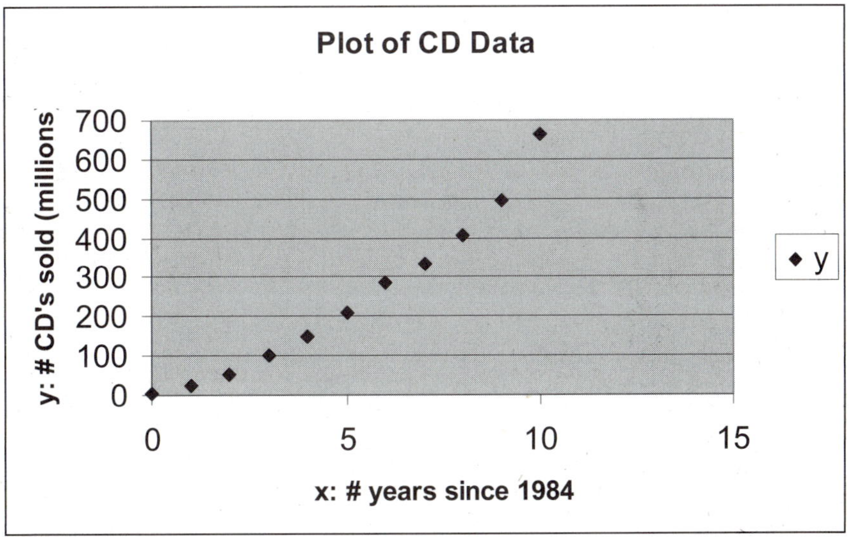

Part B: Steps to find best fit quadratic

1 Single click into the chart in your workbook.

2 From the **Chart** menu, choose Add Trendline. You will then see a dialog box like the one below. Click on the Polynomial option for Trend/Regression type. Set the Order to 2 for a quadratic. This is illustrated in the dialog box below.

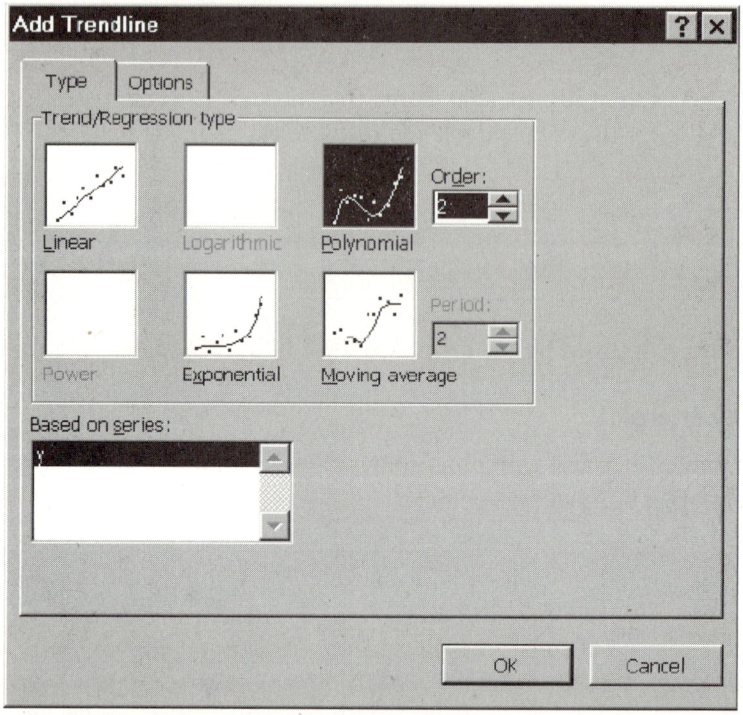

3 Click into the Options tab. Check the box for displaying equation on chart to see a dialog box like the one below.

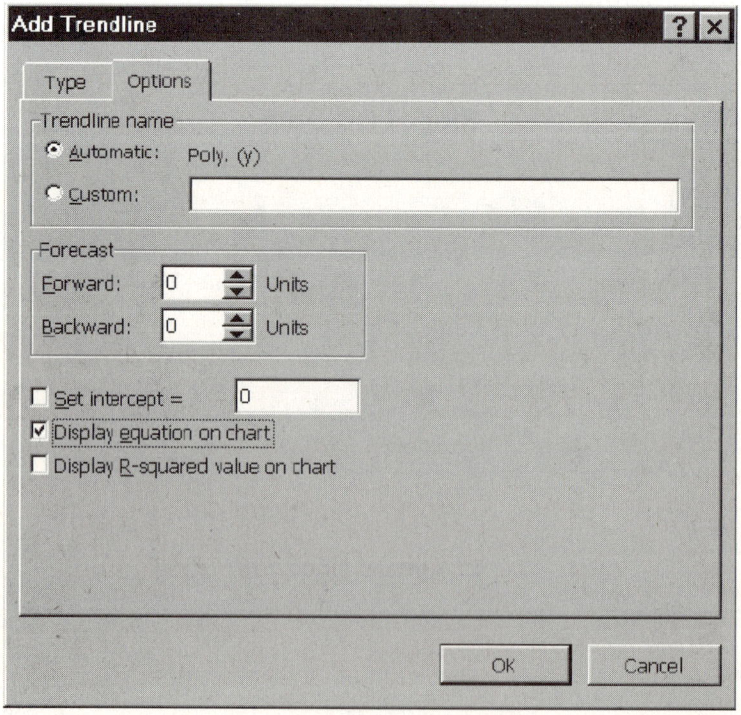

4 Click OK and you will see the best fit quadratic along with its equation as follows.

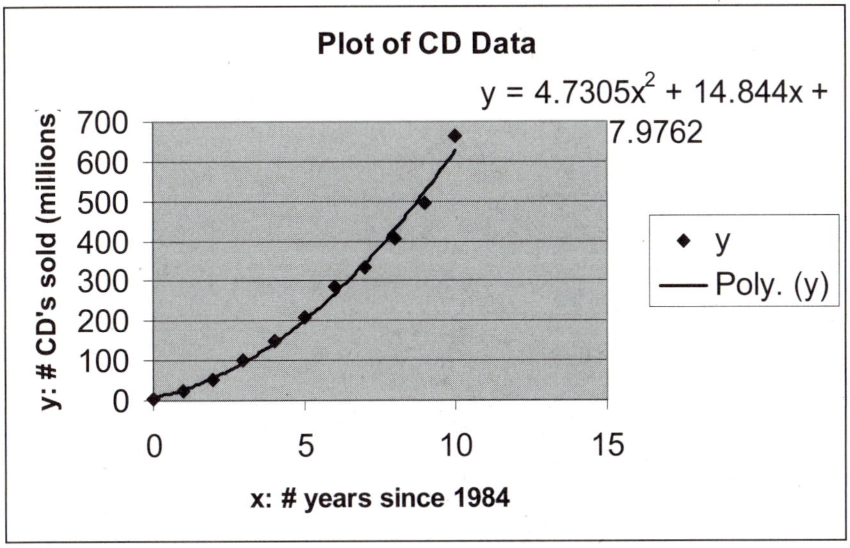

5 The best fit quadratic is therefore given by $y = 4.7305x^2 + 14.844x + 7.9762$.

6 Notes: You may have to move the equation text in your chart to a place where it is easier to see. Also, since this example uses x=0 starting from 1984, rather than 1990 as in the Larson Edwards text, the equations will not be the same, but will yield the same projections.

Part C: Projecting number of CD's in 1998

1 It is not simple to automatically output the coefficients of the quadratic in a manner comparable to the slope and intercept functions for linear regression. To simplify the discussion, type the coefficients of the quadratic in cells **`b5:b7`**, with headings in **`a5:a7`** and **`a4:b4`** as follows.

	A	B
4	Coefficients	Value
5	a	4.7305
6	b	14.844
7	c	7.9762

2 In cell **`a9`**, type the heading "x: corresponding to 1998" and in cell **`a10`**, type in 14 (why?).

3 In cell **`b9`**, type the heading "millions of CD's" and in cell **`b10`**, type the formula for the quadratic expression **`=$b$5*a10^2+$b$6*a10+$b$7`**

	A	B
9	x: corresponding to 1998	Millions of Cd's
10	14	`=$b$5*a10^2+$b$6*a10+$b$7`

4 When you press <ENTER>, your answer will be approximately 1143 in the **`B10`** cell. This means that approximately 1.143 billion CD's will be sold in 1998.

Check it out

- Compare the values predicted by the quadratic function with the actual data for the years 1984-1994 as shown in the linear regression section. From your figure and this calculation, discuss how well the quadratic approximates the actual data.

Suggested Exercises

Text	Section(s)	Exercises
Harshbarger-Reynolds *Mathematical Applications,* 7th ed.	2.5	#9-36
Larson-Edwards *Calculus: An Applied Approach,* 6th ed. (and brief version)	7.7	#9-43
Berresford-Rockett *Applied Calculus,* 3rd ed. (and brief version)	7.4	#9-18, #27-32
Berresford-Rockett *Calculus with Finite Mathematics*	11.4	#9-18, #27-31
Berresford-Rockett *Finite Mathematics*	Various	Used in modeling exercises in various sections

The techniques in this chapter can also be used for modeling exercises appearing in various chapters in the textbooks.

Chapter 4

Finding Zeros of a Function with Goal Seek

Finding the X-intercept of a Line

To find the x-value where a function is zero, you can use a feature of Excel called Goal Seek. The next example will show how to use Goal Seek.

Example Let the profit function for a company be given by $p(x) = 200x - 4000$, where x denotes the number of items produced. The manufacturer wants to know how many items to produce to break even. That is, she wants to know when the profit will be zero.

Solution The steps to solving this problem using Goal Seek are given below.

Steps for using Goal Seek to find x-intercept

1 First make a table with x and the formula for p(x):

	A	B
1	x	p(x)
2	1	=200*A2-4000

2 Change the value of x in the cell **A2** and press<Enter>. Note what happens to the value of p(x). We want to find the value of x such that p(x) = 0. Since this is a linear equation, there will be only one such value.

3 Go to **Tools > Goal Seek**. You will get the following dialog box.

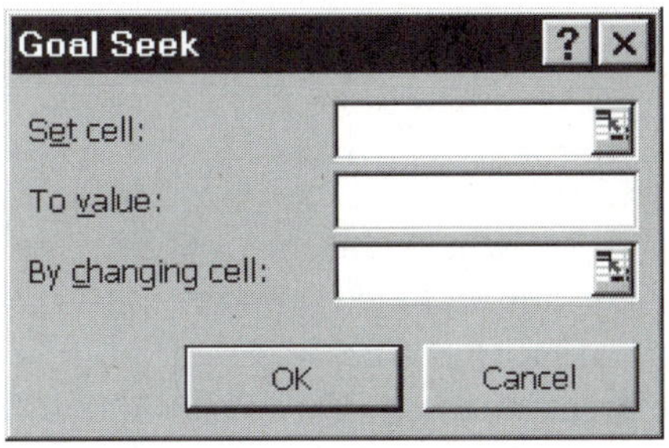

4 Click cursor into the **Set Cell** box, and click into the cell **B2** (representing profit). The dialog box will automatically record its cell reference. In the **To Value** box, type 0. Hence, you should see the following dialog box.

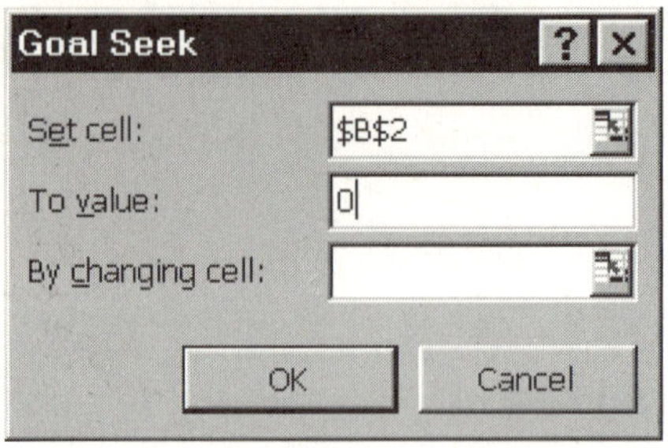

5 You complete the data entry by filling in the last box called **By changing cell**. This is the x-value. Click into the cell **A2**, and the dialog box will automatically record its cell reference. Your completed box should look like the following:

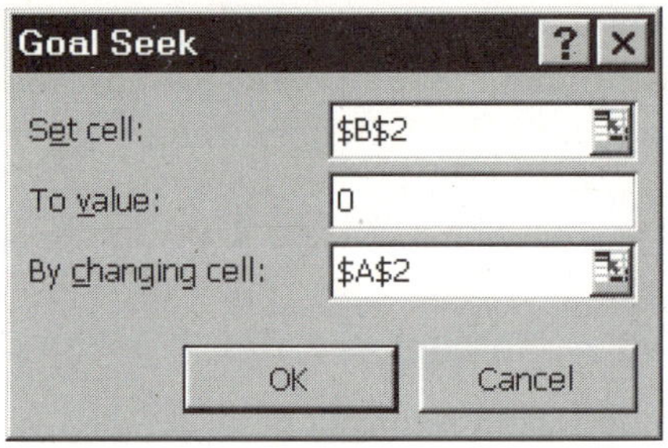

6 Click OK. Goal Seek will give you the following final dialog box.

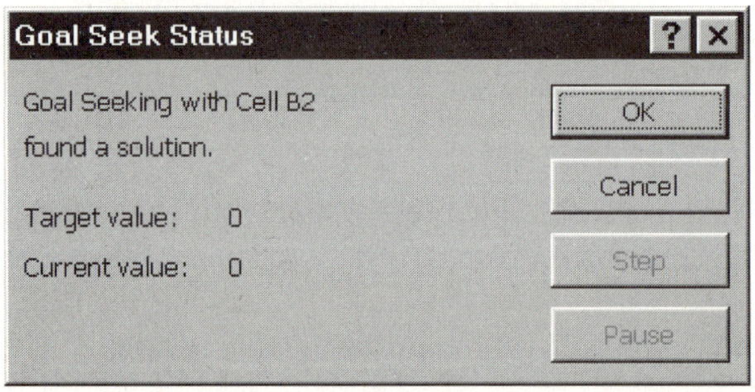

7 Click OK and the cell values in the A2 and B2 cells for x and p(x) will be changed accordingly.

Check it out

- Check the cells A2 and B2 to see what solution Goal Seek gave you. You should get a value of x=20 to make p(x)=0. This means that the company must make at least 20 products before realizing a positive amount of profit.
- Use Goal Seek to find the break-even point if p(x) = 300x-8800.

Finding Zeros of a Quadratic Function

You know from algebra that a parabola could have 0,1 or 2 x-intercepts. Goal Seek can return only one x-intercept at a time. Which one it returns depends on the value of x which is already in the box when you start Goal Seek. In the previous example, we knew there would only be one x-intercept, since the function was linear, and it did not matter what value x had when starting Goal Seek.

It is therefore advisable to graph the function before starting Goal Seek. You can then set the initial value for x close to the x-intercept you are interested in. We will illustrate this in the next example.

Example Find the zeros of the function $f(x) = x^2 - 6x + 7$.

Solution Follow the steps below to find one of the zeros of *f(x)*.

Steps to find one zero of a quadratic function

1 The vertex of the parabola is at x=-b/2a=3. Therefore, we pick an interval of x-values around x=3. For this example,we choose the interval [0,6]

2 Make a table of x and y values using the directions for tables in Chapter 1, using formulas to generate the f(x) values.

	A	B
1	x	f(x)
2	0	7
3	1	2
4	2	-1
5	3	-2
6	4	-1
7	5	2
8	6	7

3 Select the range of cells **A1:B8** and graph using the directions in Chapter 2. Your graph should look like the following:

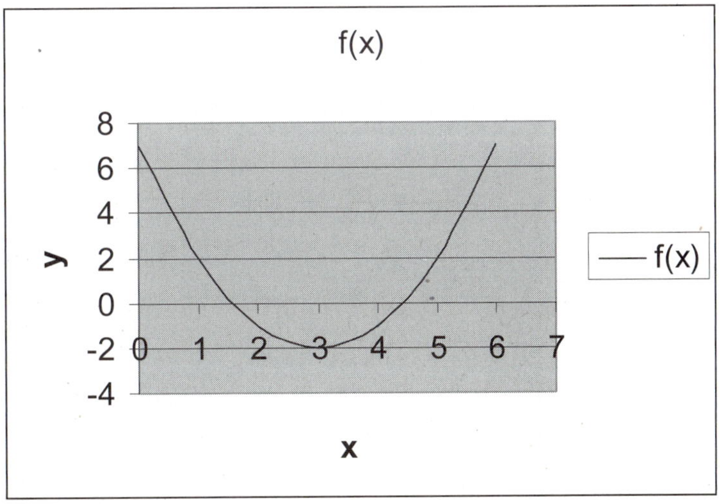

4 We see that there is one x-intercept near 2 and another near 4. We can start Goal Seek in the following table with the starting value of x=2 in **`a12`**. Formula for f(x) is entered in **`b12`**.

	A	B
11	x	f(x)
12	2	`=a12^2-6*a12+7`

5 Invoke Goal Seek from **Tools > Goal Seek**, and follow the directions given in the previous example. The box should look like the following after you entered all pertinent data.

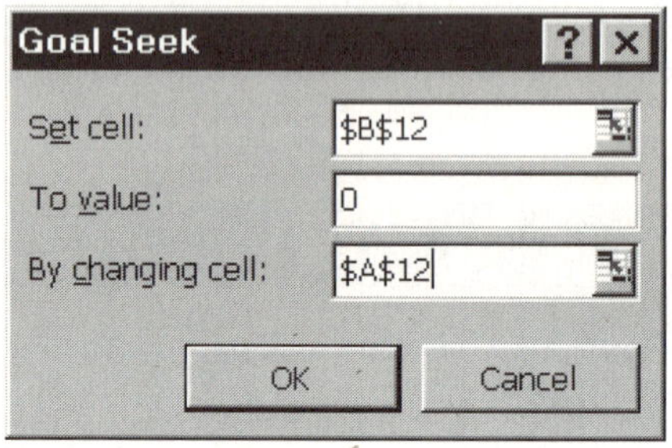

6 Click OK and you should get the following box.

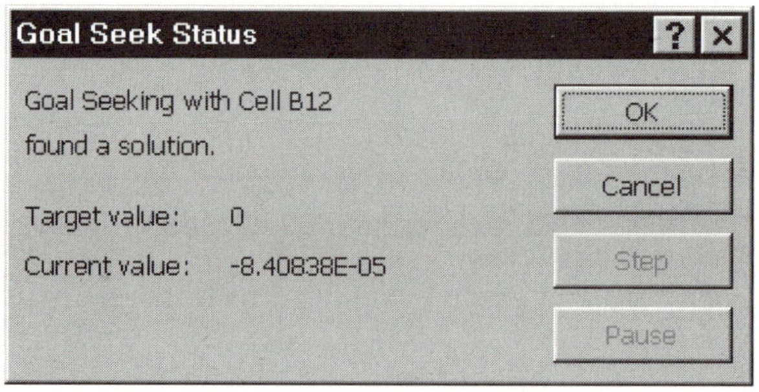

7 The x-intercept near 2 is approximately 1.585816,as illustrated in the screenshot below. Note that Goal Seek gives an approximate answer. The y-value is very small but not quite zero.

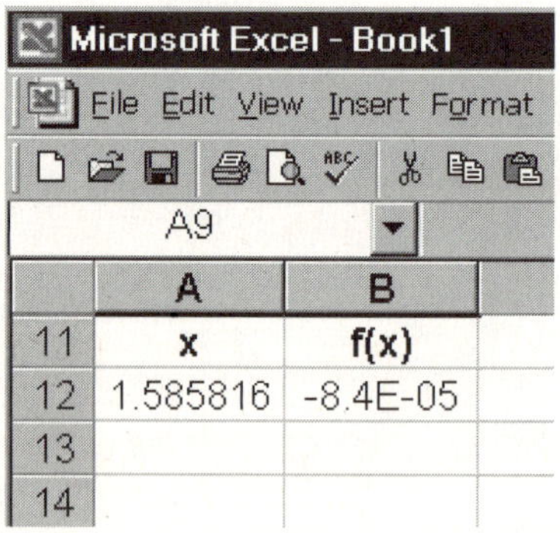

Check it out

- Find the x-intercept near 4 using Goal Seek. Your answer should be approximately 4.414.

Suggested Exercises

Text	Section(s)	Exercises
Harshbarger-Reynolds *Mathematical Applications, 7th ed.*	1.4	#35-38
Larson-Edwards *Calculus: An Applied Approach, 6th ed.* (and brief version)	1.2	#23-38
Berresford-Rockett *Applied Calculus, 3rd ed.* (and brief version)	1.3 1.4	#53-62 #21-32
Berresford-Rockett *Calculus with Finite Mathematics*	0.3 0.4	#53-62 #21-32
Berresford-Rockett *Finite Mathematics*	1.3 1.4	#53-62 #21-32

Break-even Problems using Goal Seek

Break even problems are those which require you to find a point where two quantities are equal. Examples are cost-revenue or supply-demand problems. You can use Goal Seek to find break-even points for such problems. The following example shows how.

Example The supply function for widgets is given by p=4q+1, where q is the quantity supplied and p is the price. The demand function for widgets is given by p = -3q+36. Find the equilibrium price for the widgets. (Harshbarger-Reynolds,Example 5, Section 1.6)

Solution The equilibrium price is the price for which supply equals demand. Similar to the previous examples, we must set up a table with entries for price, supply, and demand. Enter the appropriate formulas for the supply and demand, as indicated in the table below, and press <ENTER>.

	A	B	C	D
1	q	p: supply	p: demand	
2	1	**=4*A2+1**	**=-3*A2+36**	

Goal Seek will not let you equate the supply cell to the demand cell. (Try it and see what happens.) Therefore, to calculate the equilibrium price, you must make another entry with the heading supply-demand. Your table will now look like the one below.

	A	B	C	D
1	q	p: supply	p: demand	supply-demand
2	1	**=4*A2+1**	**=-3*A2+36**	**=B2-C2**

At equilibrium, supply price =demand price, which is equivalent to writing supply-demand=0. We can therefore ask Goal Seek to find the quantity for which supply-demand =0.

Steps for using Goal Seek to find the equilibrium point

1. Invoke **Tools > Goal Seek**
2. In the dialog box, set the **`Set Cell`** reference to **D2** (supply-demand)
3. Then, set the **`To Value`** to 0 (for equilibrium)
4. Finally, set **`By Changing Cell`** to **A2** (quantity is the variable that is changed)

Your dialog box looks like the following

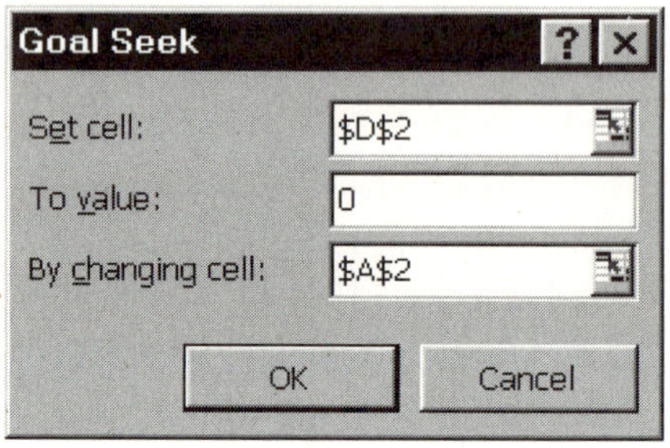

5. Click OK and you will see the solution dialog box. Click OK again. Your original cells will be changed to reflect the equilibrium quantity: q=5 and p=21.

Check it out

- Solve this problem by hand and check to see if you get the same answer.
- How many widgets are demanded at the equilibrium price? How many widgets are supplied at the equilibrium price?

Suggested Exercises

Text	Section(s)	Exercises
Harshbarger-Reynolds *Mathematical Applications, 7th ed.*	1.5 1.6	#21-24 #15-26, 31-36, 42-48
Larson-Edwards *Calculus: An Applied Approach, 6th ed.* (and brief version)	1.2 1.3	#55-68, 71-82 #71-100
Berresford-Rockett *Applied Calculus, 3rd ed.* (and brief version)	1.3	#79-82
Berresford-Rockett *Calculus with Finite Mathematics*	0.3	#79-84
Berresford-Rockett *Finite Mathematics*	1.3	#79-84

Chapter 5

Matrices

Excel can be used to add, subtract, multiply and compute inverses of matrices. To enter a matrix, simply enter each of the elements of the matrix in a cell. To manipulate the matrices, formulas are used which work on the entire matrix. For this reason, these formulas are called array formulas.

Adding and Multiplying Matrices

Example Add the matrices A= $\begin{bmatrix} 1 & 2 & 3 \\ 4 & 0 & 1 \end{bmatrix}$ and B= $\begin{bmatrix} 3 & 5 & -2 \\ -3 & 2 & 3 \end{bmatrix}$.

Solution

Steps for adding matrices

1 Create heading titled "A" in cell `a1` for the first matrix .
2 Enter each matrix element of A in each cell from `b1:d2` . See the figure below.
3 Similarly, type a heading for matrix B in cell `a4` .
4 Enter the matrix B in cells `b4:d5` .
5 Since both matrices are the same size, we can add them together, element by element.
6 Type a heading "A+B" in cell `a7` .
7 In cell b7, type the formula `=b1+b4` .
8 Copy this formula across the row to `c7:d7`
9 Select the row `b7:d7`
10 Copy the entire row `b7:d7` to `b8:d8`
11 Your spreadsheet should look like the following (shown in formula view).

	A	B	C	D
1	**A**	1	2	3
2		4	0	1
3				
4	**B**	3	5	-2
5		-3	2	3
6				
7	**A+B**	=B1+B4	=C1+C4	=D1+D4
8		=B2+B5	=C2+C5	=D2+D5

Check it out

- With A and B as above, find A-B and 2A+B. Remember to recopy your formulas.

Multiplication of Matrices

To multiply matrices in Excel, you use a function called MMULT, which takes the cell ranges of two matrices as its argument.

Example Find AB where A= $\begin{bmatrix} 1 & -2 \\ 0 & 4 \\ -1 & 0 \end{bmatrix}$ and B= $\begin{bmatrix} -1 & 2 \\ 0 & 1 \end{bmatrix}$.

Solution

Steps for multiplying matrices

1. Since A is 3x2 and B is 2x2, the multiplication is defined since the number of columns in A equals the number of rows in B.
2. Enter the heading "A" in cell **`a1`**, and the matrix A in cells **`b1:c3`** .
3. Enter the heading "B" in cell **`a5`**, and the matrix B in cells **`b5:c6`** .
4. Enter the heading "AxB" in cell **`a8`**
5. The product will be of size 3x2. Therefore select a range of cells of this size where the product will appear - for example, the range **`b8:c10`**.

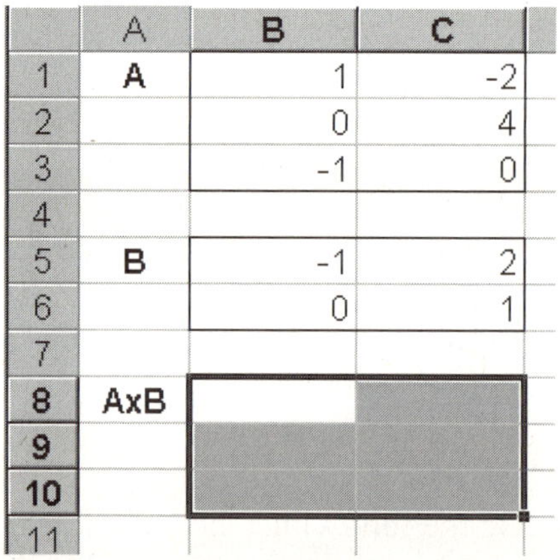

6. In the formula bar, type **`=mmult(`** and then select the matrix A. You will see the following on your spreadsheet.

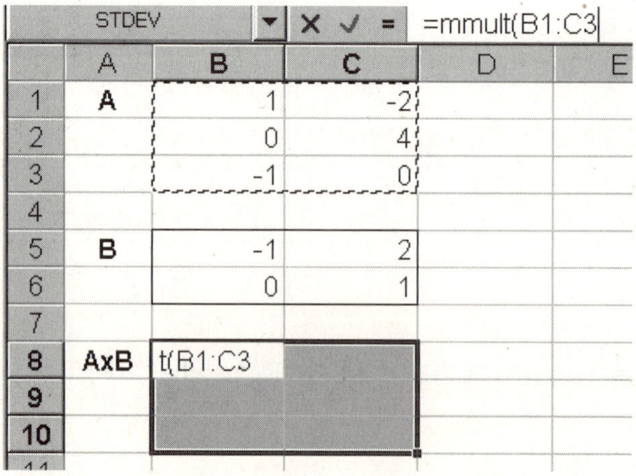

7 Staying in the formula bar type a comma and then select matrix B. You will see the following on your spreadsheet.

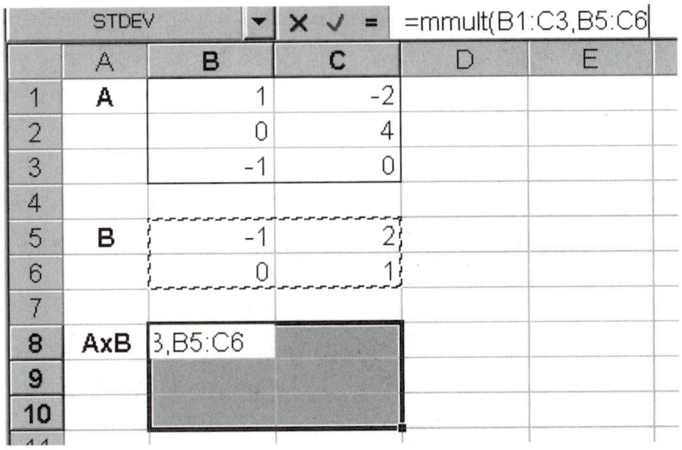

8 Finish the formula in the formula bar by typing the right parentheses and then press <CTRL><SHIFT><ENTER> **all at the same time**. Your screen will be similar to the one below. The computer will automatically insert the braces since this is an array formula.

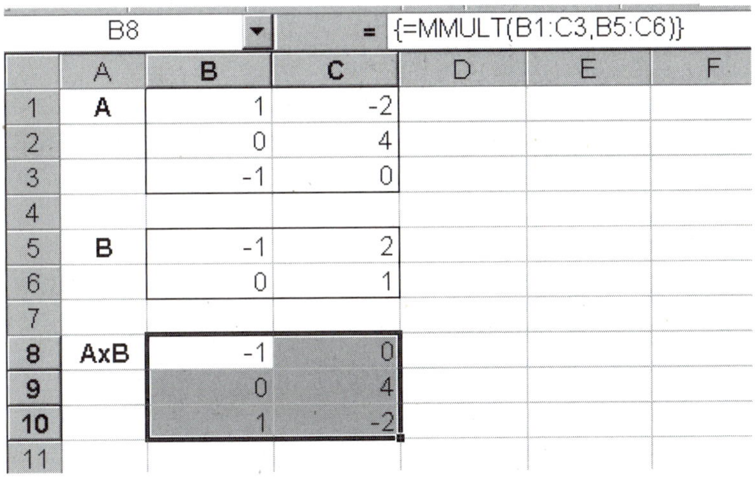

Note: You must press <CTRL><SHIFT><ENTER> **all at the same time after entering the array formula**. Otherwise, only one of the matrix elements will appear.

Check it out

- Change some numbers in A or B, press <ENTER> and see what happens to the product.
- Add another column of numbers to B to make it a 2x3 matrix and compute BA.

Suggested Exercises

Text	Section(s)	Exercises
Harshbarger-Reynolds *Mathematical Applications, 7th ed.*	3.2	#53,54,57,58,62-71
Berresford-Rockett *Calculus with Finite Mathematics, 6th ed.*	2.4	#21-30, #51-60
Berresford-Rockett *Finite Mathematics*	3.4 7.1,7.2	#21-30, #51-60 All exercises with calculator icon

Inverse of a Matrix

Example Find the inverse of the matrix $\begin{bmatrix} 2 & 5 & 4 \\ 1 & 4 & 3 \\ 1 & -3 & -2 \end{bmatrix}$ (Ex. 2, Sec. 3.4, Harshbarger-Reynolds).

Solution

Steps for finding the inverse

1 Enter the heading “A” in cell **`a1`**, and the matrix A in cells **`b1:d3`** .
2 Enter the heading “inverse (A)” in cell **`a5`** .
3 Select the range of cells where the inverse should appear. Since the inverse of the matrix will be of the same size, select a 3x3 region, for example **`b5:d7`** .

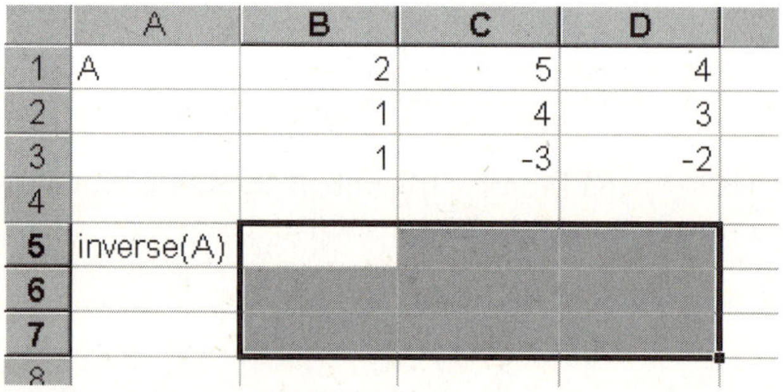

	A	B	C	D
1	A	2	5	4
2		1	4	3
3		1	-3	-2
4				
5	inverse(A)			
6				
7				
8				

4 In the formula bar, enter the formula **`=minverse(`**

5 Select the matrix A.

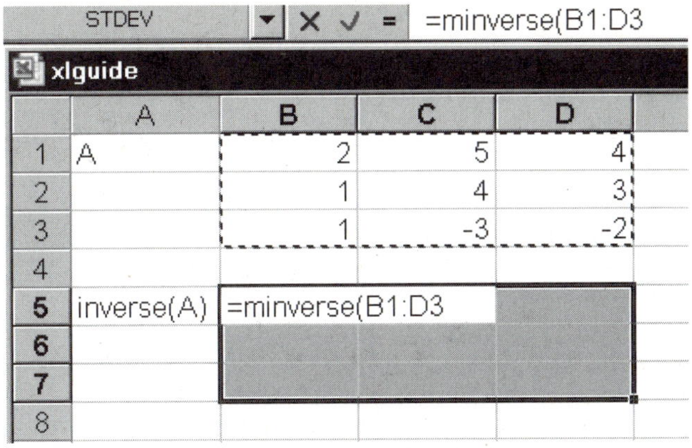

6 Close the parentheses in the formula and press <CTRL><SHIFT><ENTER> **all at the same time**. Your screen will be similar to the one below.

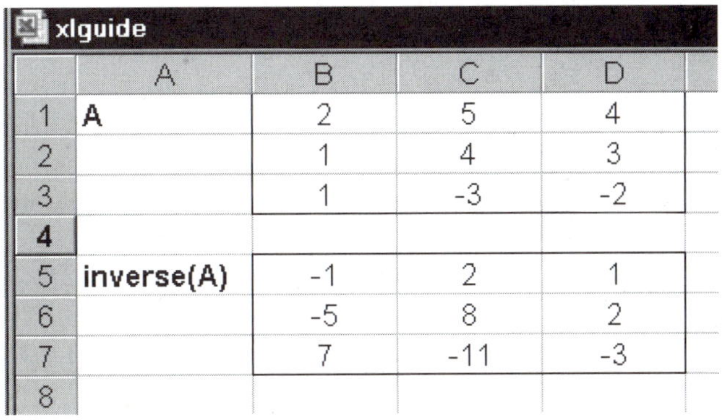

Suggested Exercises

Text	Section(s)	Exercises
Harshbarger-Reynolds *Mathematical Applications, 7th ed.*	3.4	#23,24;#35-54
Berresford-Rockett *Calculus with Finite Mathematics, 6th ed.*	2.5	#9,10,29,30
Berresford-Rockett *Finite Mathematics*	3.5 7.3	#9,10,29,30 All exercises with calculator icon

Solving Systems of Equations using Inverses

Matrix inverses can be used for solutions of linear systems of equations. The next example shows how this is accomplished in the spreadsheet.

Example Solve the system of equations

$$\begin{cases} 2x + 5y + 4z = 4 \\ x + 4y + 3z = 1 \\ x - 3y - 2z = 5 \end{cases}$$

Solution The corresponding matrix equation is

$$\begin{bmatrix} 2 & 5 & 4 \\ 1 & 4 & 3 \\ 1 & -3 & -2 \end{bmatrix} \begin{bmatrix} x \\ y \\ z \end{bmatrix} = \begin{bmatrix} 4 \\ 1 \\ 5 \end{bmatrix}$$

Its solution is given by

$$X = \begin{bmatrix} x \\ y \\ z \end{bmatrix} = \begin{bmatrix} 2 & 5 & 4 \\ 1 & 4 & 3 \\ 1 & -3 & -2 \end{bmatrix}^{-1} \begin{bmatrix} 4 \\ 1 \\ 5 \end{bmatrix}$$

Steps to compute the solution in Excel

1 Since we need to find the inverse of the same matrix as in the example on matrix inverses, repeat Steps (1)-(6) on how to find the matrix inverse.

2 Enter the 3x1 matrix B= $\begin{bmatrix} 4 \\ 1 \\ 5 \end{bmatrix}$ in cells **`b9:b11`**, with a heading "B" in **`a9`** .

3 Enter a heading "X" for the solution in cell a13.

4 The solution X is given by the product $A^{-1}B$. This product will be computed using MMULT in the cells **`b13:b15`** .

a Select the cells **`b13:b15`** .

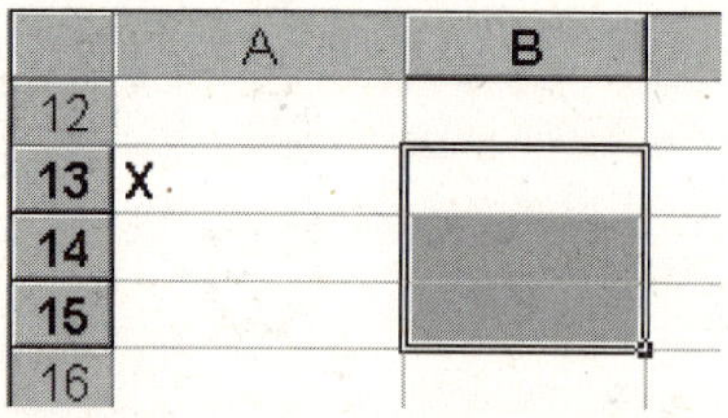

b In the formula bar, type **`=mmult(`**

c Select the matrix A^{-1}

FV ▾ ✕ ✓ = =mmult(B5:D7

	A	B	C	D
1	A	2	5	4
2		1	4	3
3		1	-3	-2
4				
5	inverse(A)	-1	2	1
6		-5	8	2
7		7	-11	-3
8				
9	B	4		
10		1		
11		5		
12				
13	X	=mmult(B5:D7		
14				
15				

d Type a comma and select the matrix B.

e Type the closing parentheses and press <CTRL><SHIFT><ENTER> **all at the same time**. You result will be as follows.

	A	B	C	D
1	A	2	5	4
2		1	4	3
3		1	-3	-2
4				
5	inverse(A)	-1	2	1
6		-5	8	2
7		7	-11	-3
8				
9	B	4		
10		1		
11		5		
12				
13	X	3		
14		-2		
15		2		

f Hence, the solution to the system of equations is x=3,y=-2, and z=2.

Check it out

- Check that the solution given above actually works.Change the value of B to other set of numbers.
- Compute and check your new solution.

Suggested Exercises

Text	Section(s)	Exercises
Harshbarger-Reynolds *Mathematical Applications, 7th ed.*	3.4	#35-42
Berresford-Rockett *Calculus with Finite Mathematics, 6th ed.*	2.5	#45-60
Berresford-Rockett *Finite Mathematics*	3.5	#45-60

Leontief Input-Output Model

The material in the preceding sections can be easily used to implement the calculations for the Leontief Input-Output Model. Therefore, we will only outline the necessary steps involved.

To solve the matrix equation $(I-A)^{-1}X=D$, follow these steps:

1. Form the matrix (I-A).
2. Find $(I-A)^{-1}$ using the method shown in the matrix inverse section.
3. Find X by multiplying $(I-A)^{-1}$ by D, as illustrated in the section on solving linear systems of equations.

Chapter 6

Linear Programming using Solver

This chapter will illustrate the use of an Excel tool called Solver to solve linear programming problems. To check that your installation of Excel has Solver, select Tools in the menu bar. If you see Solver as one of the options, you are ready to go. Otherwise, you will have to add it in. See the preface on how to add in Solver.

Maximization Problem using Solver

Example 1 (Example 3, Sec. 4.4, Harshbarger -Reynolds)

The Solar Technology Company manufactures three different types of hand calculators and classifies them as small, medium, and large according to their calculating capabilities. The three types have production requirements given by the following table:

	Small	**Medium**	**Large**
Electronic circuit components	5	7	10
Assembly time (hours)	1	3	4
Cases	1	1	1

The firm has a monthly limit of 90000 circuit components, 30000 hours of labor, and 9000 cases. If the profit is \$6 for the small, \$13 for the medium, and \$20 for the large calculators, how many of each should be produced to yield maximum profit?

Solution

Set up of problem

1 Identify variables

x: number of small calculators

y: number of medium calculators

z: number of large calculators

2 Identify objective: Maximize the profit function $f = 6x + 13y + 20z$

3 The objective function is subject to the following constraints:

$$5x + 7y + 10z \leq 90000$$
$$x + 3y + 4z \leq 30000$$
$$x + y + z \leq 9000$$
$$x, y, z \geq 0$$

The next step is to input all this information into Excel so that Solver can be invoked. Since all the calculations in the spreadsheet are done with cell references, you must set up cell entries for the variables, objective function and constraints. In Excel, the cell containing the formula for the objective function is referred to as the *target cell.* The cells containing the variables are called the *changing cells.* The constraints are referred to as, well, *constraints*.

Steps to set up the problem in Excel

1 In a blank spreadsheet, first type a heading called "Variables" in cell **a1**, followed by the variable descriptions in **a3:a5** and values in cells **b3:b5**. The variables are initially assigned values of zero. Refer to the table below as a guide.

2 The objective function formula is given in terms of the cell references for the variables x,y, and z. Enter the information for the objective function as follows:

- **a** Type a heading called "Objective" in cell **a7**
- **b** Type a description of the objective in cell **a9**
- **c** Enter the objective function formula in **b9** .The formula is **=6*b3+13*b4+20*b5**

3 Type in the formulas for the constraints.

- **a** Type a heading called "Constraints" in **a11** and descriptive labels in **a13:a15**.
- **b** The formulas for the constraints are also given in terms of the cell references for x,y, and z and are contained in **b13:b15** .
- **c** The maximum available is typed in **c13:c15**. The complete setup of formulas and other entries is shown below.

	A	B	C
1	**Variables**		
2			
3	# small calculators (x)	0	
4	# medium calculators (y)	0	
5	# large calculators (z)	0	
6			
7	**Objective**		
8			
9	Maximize profit	=6*B3+13*B4+20*B5	
10			
11	**Constraints**		
12		Amount used	Maximum
13	Circuit components	=5*B3+7*B4+10*B5	90000
14	Labor	=B3+3*B4+4*B5	30000
15	Cases	=B3+B4+B5	9000

Check it out

- To get familiar with the setup of the problem in Excel, change the variables in `b3:b5` to some nonzero values. What happens to the value of the objective function? What happens to the values for the constraints?

Steps to solve the problem using Solver

1. Once you check that your spreadsheet contains all the correct formulas in the appropriate cells, you are ready to invoke Solver by choosing **Tools > Solver**.
2. You will see a dialog box whose first entry is the information for the target cell (i.e. the objective function). Click cursor into the this entry box and click into cell `B9` (formula for objective function).
3. Check the button to maximize. Your dialog box should now look like the one below.

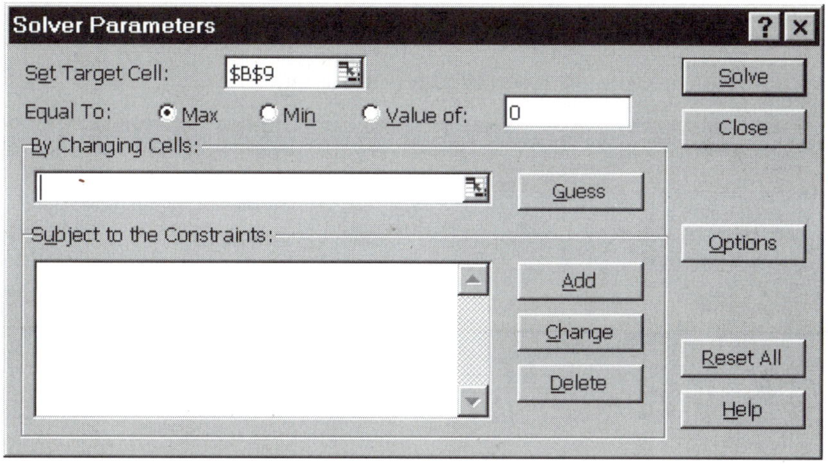

4. Next Click cursor to the By Changing Cells entry box.
5. Enter the cell references for the variables by selecting the cells `b3:b5` . This portion of your dialog box should resemble the following.

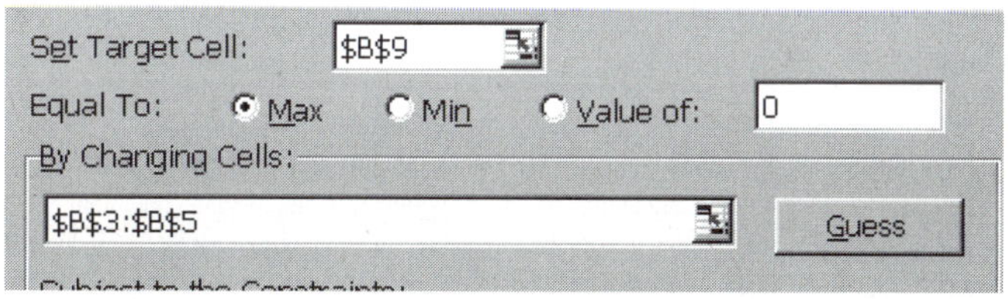

6. Now you will add the constraints.
   - **a** Click cursor into Subject to the Constraints entry box.
   - **b** Press the Add button to add the first constraint. You will get a new dialog box for the constraint.
   - **c** Click cursor to the left entry box and click into cell `b13` containing the formula for the first constraint.
   - **d** The middle entry box should be set to <=.

e Click cursor to the right entry box and click into the cell c13 containing the maximum quantity. Your constraint dialog box should resemble the following.

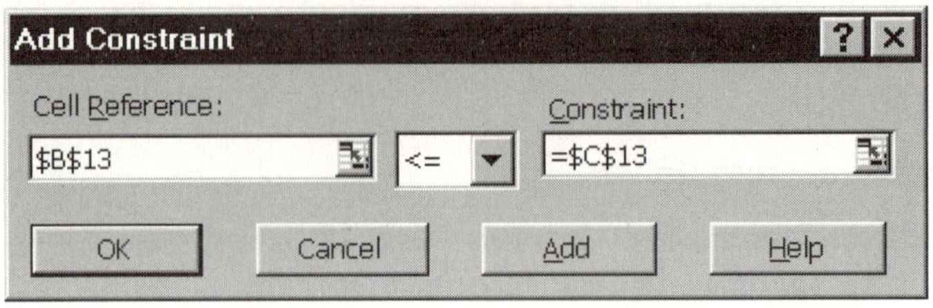

f Click the Add button to add the labor constraint and repeat Steps (c)-(e).

g Click the Add button and repeat Steps (c)-(e) for the cases constraint.

h Now add the nonnegativity constraints. Click into the left entry box for the constraint and select the variables in cells `b3:b5`. Set the middle entry box to >=. Type 0 into the right entry box.

i Click OK.

7 Your completed Solver box should resemble the following.

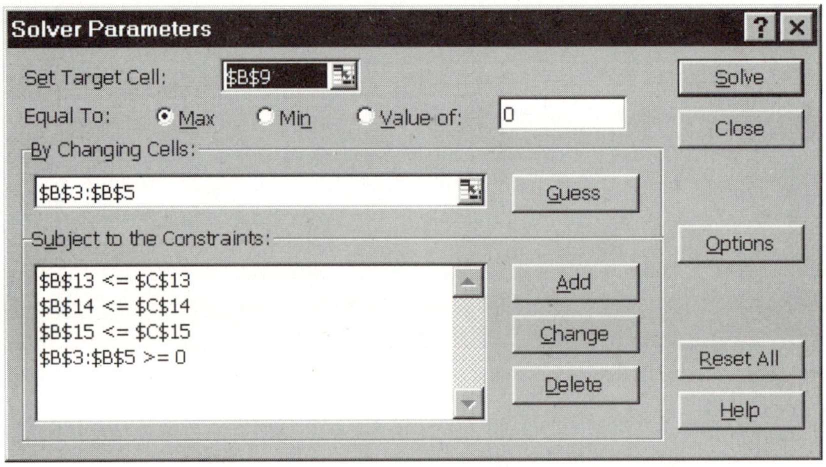

8 Click into the Options box, and make sure that the Assume Linear Model checkbox is checked, as in the following figure.

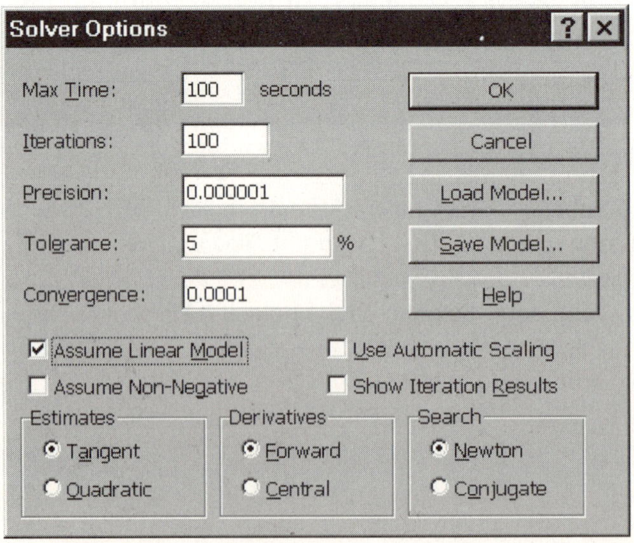

9 Click OK in the Options dialog box.

10 Click Solve in the Solver dialog box. You will get a dialog box stating that Solver found a solution.

11 Check the Keep Solver Solution button and also select the Answer report. You completed box resembles the following

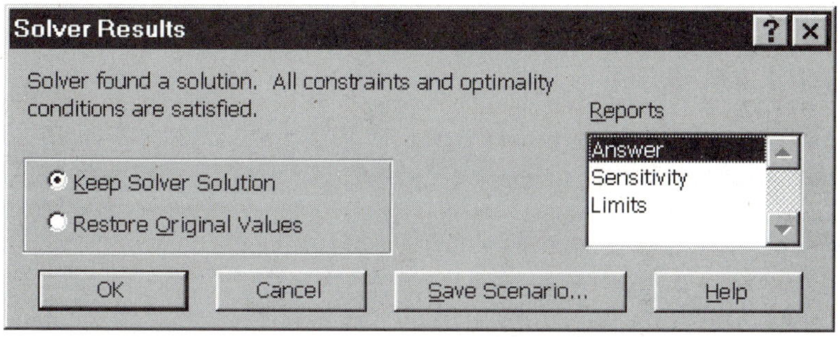

12 Click OK. Go back and examine the cells with the variables, constraints, and objective. They should now contain the optimal values and resemble the following table.

Variables

# small calculators (x)	2000
# medium calculators (y)	0
# large calculators (z)	7000

Objective

Maximize profit	152000

Constraints

	Amount used	Maximum
Circuit components	80000	90000
Labor	30000	30000
Cases	9000	9000

13 Click on the worksheet labeled Answer Report 1 to see a summary of the solution.

Suggested Exercises

Text	Section(s)	Exercises
Harshbarger-Reynolds *Mathematical Applications, 7th ed.*	4.4	#47-59
Berresford-Rockett *Calculus with Finite Mathematics*	3.3	#41-50
Berresford-Rockett *Finite Mathematics*	4.3	#41-50

Minimization Problem using Solver

In Solver, minimization problems and problems with mixed constraints are handled in a manner entirely similar to the above example. For completeness, the next example is a minimization problem.

Example 2 (Example 3, Sec. 4.5, Harshbarger-Reynolds)

A beef producer is considering two different types of feed. Each feed contains some or all of the necessary ingredients for fattening beef. Brand 1 feed costs 20 cents per pound and brand 2 costs 30 cents per pound. How much of each brand should the producer buy in order to satisfy the nutritional requirements for Ingredients A and B at minimum cost? The following tables contains the relevant information about nutritional requirements and cost.

	Brand 1	Brand 2	Minimum Requirement
Ingredient A	3 units/lb	5 units/lb	40 units
Ingredient B	4 units/lb	3 units/lb	46 units
Cost per pound	20 cents	30 cents	

Solution From the textbook, the setup for this problem is as follows

Set up of problem

1 Identify variables:

x: pounds of Brand 1 feed

y: pounds of Brand 2 feed

2 Identify the objective function:

$$\text{Minimize } C = 20x + 30y$$

3 Identify the constraints:

$$3x + 5y \geq 40$$
$$4x + 3y \geq 46$$
$$x \geq 0,\ y \geq 0$$

Steps to set up the problem in Excel

Proceed as in Steps 1-3 in the "Steps to set up problem in Excel" section of the previous example. You will need to adjust the number of variables and type in different formulas for the objective and constraint, of course. Your complete setup should be similar to the following.

	A	B	C
1	**Variables**		
2			
3	Pounds Brand 1	0	
4	Pounds Brand 2	0	
5			
6	**Objective**		
7			
8	Minimize cost	=20*B3+30*B4	
9			
10	**Constraints**		
11		Amount in feeds	Required
12	Ingredient A	=3*B3+5*B4	40
13	Ingredient B	=4*B3+3*B4	46
14			

Check it out

- To get familiar with the setup of the problem in Excel, change the variables in **`b3:b4`** to some nonzero values. What happens to the value of the objective function? What happens to the values for the constraints?

Steps to solve the problem using Solver

1 Follow Steps 1-3 in the "Steps to solve the problem using Solver" section of the previous example, adjusting for the cell references for this example. Click the option to **minimize (Min)**.
2 Next click cursor to the By Changing Cells entry box.
3 Enter the cell references for the variables by selecting the cells **`b3:b4`**.
4 Now you will add the constraints.

  a Click cursor into Subject to the Constraints entry box.
  b Press the Add button to add the first constraint. You will get a new dialog box for the constraint.

- **c** Click cursor to the left entry box and click into cell **b12** containing the formula for the first constraint.
- **d** The middle entry box should be set to >=.
- **e** Click cursor to the right entry box and click into the cell **c12** containing the maximum quantity.
- **f** Click the Add button to add the second constraint.
- **g** Click cursor to the left entry box and click into cell **b13** containing the formula for the second constraint.
- **h** The middle entry box should be set to >=.
- **i** Click cursor to the right entry box and click into the cell **c13** containing the second nutritional constraint.
- **j** Now add the nonnegativity constraints. Click into the left entry box for the constraint and select the variables in cells **b3:b4**. Set the middle entry box to >=. Type 0 into the right entry box.
- **k** Click OK.

5 Your completed Solver box should resemble the following.

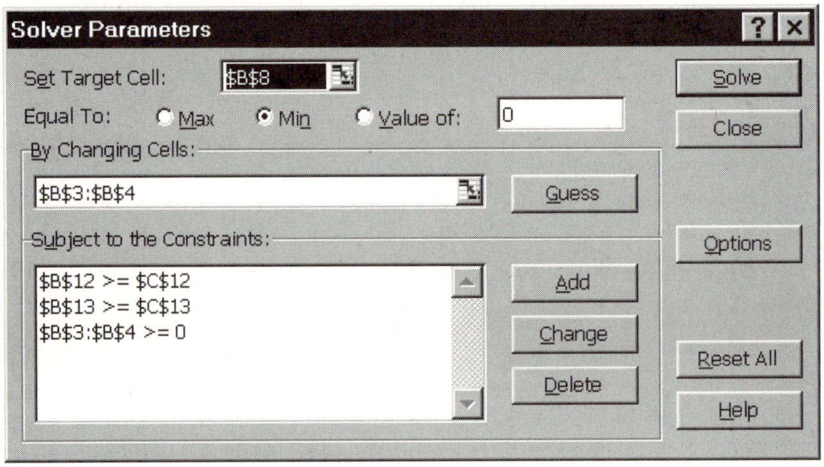

6 Click into the Options box, and make sure that the Assume Linear Model checkbox is checked. Click OK in the Options dialog box.

7 Click Solve in the Solver dialog box. You will get a dialog box stating that Solver found a solution.

8 Check the Keep Solver Solution button and also select the Answer report. Click OK. Go back and examine the cells with the variables, constraints, and objective. They should now contain the optimal values and resemble the following table.

	A	B	C
1	**Variables**		
2			
3	Pounds Brand	10	
4	Pounds Brand	2	
5			
6	**Objective**		
7			
8	Minimize cost	260	
9			
10	**Constraints**		
11		Amount in feeds	Required
12	Ingredient A	40	40
13	Ingredient B	46	46

9 From the results, the farmer should purchase 10 pounds of Brand A and 2 pounds of Brand B to minimize cost at $2.60.

10 Click on the worksheet labeled Answer Report to see a summary of the solution.

Suggested Exercises

Text	Section(s)	Exercises
Harshbarger-Reynolds	4.5	#23-34
Mathematical Applications, 7th ed.	4.6	#35-42
Berresford-Rockett	3.4	#31-40
Calculus with Finite Mathematics	3.5	#31-40
Berresford-Rockett	4.4	#31-40
Finite Mathematics	4.5	#31-40

Chapter 7

Mathematics of Finance

A spreadsheet is an excellent tool to explore the various topics in the mathematics of finance. Since spreadsheets are used widely in the business world for financial documents, Excel has several built-in financial functions. In this chapter we will introduce many of these functions as well as explore concepts of interest, loans, annuities and mortgages.

Simple Interest

The formula for simple interest is $I=Prt$, where I is the total interest, P is the principal, r is the rate and t is the time.

Example Calculate the simple interest over 5 years for $1000 earning 6% annual interest.

Solution

1 Make a table with headings and formula as shown below:

	A	B
1	Principal	Rate
2	1000	0.06
3		
4	Year	Interest
5	1	**=a2*b2*a5**
6	2	
7	3	
8	4	
9	5	

2 Since the principal and interest should not change as we copy the formula in **b5** down the column, we use absolute references for the references containing the values for the principal and interest. Absolute references are denoted by **a2** and **b2** rather than **a2** and **b2**, respectively.

3 Copy the formula in **b5** to **b6:b9**.

4 Your finished table will be similar to the one below. We see from the table that at the end of 5 years, the investment will be worth \$1000+\$300=\$1300.

	A	B
1	**Principal**	**Rate**
2	1000	0.06
3		
4	**Year**	**Interest**
5	1	60
6	2	120
7	3	180
8	4	240
9	5	300

Suggested Exercises

Text	Section(s)	Exercises
Harshbarger-Reynolds *Mathematical Applications,* 7th ed.	6.1	#1-22
Berresford-Rockett *Calculus with Finite Mathematics*	1.1	#1-45
Berresford-Rockett *Finite Mathematics*	2.1	#1-45

Compounded Interest Using Tables

Note that the simple interest formula calculates interest only on the principal initially invested. Almost all investments do NOT calculate interest this way. Rather, they compound the interest. This means that interest is calculated on the principal and interest earned up to the point where the interest is recalculated. The period of compounding tells you how often the interest is recalculated.

Example 1 Calculate the total interest earned over five years for an investment of \$1000 earning 6% annual interest compounded annually. Compare with the simple interest example in the section above.

Solution

1 Set up a table like the one below. Note that since interest is compounded annually, the rate per period is 0.06/1=0.06.

	A	B	C	D
1	Periodic Rate			
2	0.06			
3				

	A	B	C	D
4	Period	Amount	Interest	Amount+Interest
5	1	1000	`=$a$2*b5`	`=b5+c5`

2 To calculate the interest for the second year, we take the amount in d5 to be the amount that the interest is calculated on. Therefore, the formulas for the following year should be as follows:

	A	B	C	D
4	Period	Amount	Interest	Amount+Interest
5	1	1000	`=$a$2*b5`	`=b5+c5`
6	`=a5+1`	`=d5`	`=$a$2*b6`	`=b6+c6`

3 Note that the new amount in **`b6`** is the amount + interest after 1 year, calculated in **`d5`**. The calculated values are as follows.

4 We simply repeat this process for the subsequent years. Copy the formula in **`a6:d6`** down to **`a9:d9`**. Your finished table should resemble the following.

	A	B	C	D
1	**Periodic Rate**	**Rate**		
2	0.06	0.06		
3				
4	**Period**	**Amount**	**Interest**	**Amount+interest**
5	1	1000	60	1060
6	2	1060	63.6	1123.6
7	3	1123.6	67.42	1191.02
8	4	1191.02	71.46	1262.48
9	5	1262.48	75.75	1338.23

5 Note that the interest after 5 years when compounding is used is higher than simple interest after 5 years (why?)

Check it out

- Change the compounding period in the above example to semiannually. How much interest will be earned after 5 years? Note that now you will have to calculate for 10 periods since each year has two periods.

Suggested Exercises

Text	Section(s)	Exercises
Harshbarger-Reynolds *Mathematical Applications, 7th* ed.	6.2	#49,50
Berresford-Rockett *Calculus with Finite Mathematics*	1.2	#1-7
Berresford-Rockett *Finite Mathematics*	2.2	#1-7

Compounded Interest Using Excel Functions

Making tables to calculate the interest and future value of investments can be tedious, particularly if you are interested in long term investments. Excel therefore has a built-in financial function called FV which will return the future value of an investment.

Example Use the FV function to calculate the amount in an investment after five years if the principal is $1000 and the interest rate is 6% compounded annually.

Solution To use the Excel function FV, you must first know what parameters that it takes. The FV function has the following syntax: **`FV(rate,nper,pmt,pv,type)`**

`Rate` is the interest rate per period.

`Nper` is the total number of payment periods

`Pmt` is the payment made each period; in this case, it is set to 0

`Pv` is the principal value

`Type` is the number 0 or 1 and indicates when payments are due. Since there are no payments due for this problem, simply set it to 0.

1 Make a table entering all the pertinent information as follows:

	A	B	C	D	E	F
1	Principal	Number of periods	Payment	Annual Rate	Periods per year	Periodic Rate
2	1000	5	0	0.06	1	**`=d2/e2`**

2 Note that the periodic rate is calculated automatically using a formula. This gives you the flexibility to change the Number of Periods, Annual Rate and Periods per Year without having to recalculate the periodic rate.

3 Type the heading "Future Value" in cell **`A4`**.

4 In cell **`B4`**, type the formula **`=fv(f2,b2,c2,a2,0)`**

5 The cell references in the formula above correspond to the syntax for the Fv function. Make sure you understand each argument of the FV function and where it comes from.

6 Your finished table should look like the following.

	A	B	C	D	E	F
1	**Principal**	**Number of Periods**	**Payment**	**Annual Rate**	**Periods per year**	**Periodic Rate**
2	1000	5	0	0.06	1	0.06
3						
4	**Future Value**	**($1,338.23)**				

7 Note that the future value, $1338.23, is denoted in parentheses. This means that the amount is "negative" since the money is being paid out. If you want a positive amount, you must enter -1000 for the principal. In this discussion, we will not make the distinction and simply type in the amounts as given in the problem.

Check it out

- Change the compounding in the example above from annually to quarterly. You will have to modify the values for Number of Periods and Periods per Year. What is the future value for this case?

Suggested Exercises

Text	Section(s)	Exercises
Harshbarger-Reynolds *Mathematical Applications, 7th* ed.	6.2	#7-32, 65-83
Berresford-Rockett *Calculus with Finite Mathematics*	1.2	#8-90, 106-120
Berresford-Rockett *Finite Mathematics*	2.2	#8-44

Future Value of Ordinary Annuities and Annuities Due

Excel's FV function can be easily used to calculate the future value of an annuity. The following example will show you how.

Example 1 $100 is deposited at the end of month in an account which pays 8% interest compounded monthly. How much money will be in the account at the end of eighteen months?

Solution To use the Excel function FV, you must first know what parameters that it takes. The FV function has the following syntax: `FV(rate,nper,pmt,pv,type)`

`Rate` is the interest rate per period.

`Nper` is the total number of payment periods

`Pmt` is the payment made each period; in this case, it is set to 100

`Pv` is the initial deposit; in this example, it is set to 0

`Type` is the number 0 or 1 and indicates when payments are due. Set type to 0 if payments are due at the end of the period. Set type to 1 if payments are due at the beginning. For this example, type is set to 0.

1 Make a table entering all the pertinent information as follows:

	A	B	C	D	E	F
1	Pv	Number of periods	Payment	Annual Rate	Periods per year	Periodic Rate
2	0	18	100	0.08	12	`=d2/e2`

2 Note that the periodic rate is calculated automatically using a formula. This gives you the flexibility to change the Number of Periods, Annual Rate and Periods per Year without having to recalculate the periodic rate.
3 Type the heading "Future Value" in cell **`A4`**.
4 In cell **`B4`**, type the formula **`=fv(f2,b2,c2,a2,0)`**
5 The cell references in the formula above correspond to the syntax for the FV function. Make sure you understand each argument of the FV function and where it comes from.
6 Your finished table should look like the following.

	A	B	C	D	E	F
1	Pv	Number of Periods	Payment	Annual Rate	Periods per year	Periodic Rate
2	0	18	100	0.08	12	0.006667
3						
4	Future Value	($1,905.72)				

7 Hence, at the end of eighteen months, the account will be valued at $1905.72.

Check it out

- Change the annual rate to 10% in the above example, leaving everything else the same. What is the amount after eighteen months? after two years?

In problems involving **annuities due**, payments are made at the *beginning* of each period. Using Excel's FV function, this amounts to simply changing one of the parameters when calling the function.

Example 2 $150 is deposited at the beginning of each month in an account which pays 7% interest compounded monthly. How much money will be in the account at the end of eighteen months?

Solution To use the Excel function FV, you must first know what parameters that it takes. The FV function has the following syntax: **`FV(rate,nper,pmt,pv,type)`**

`Rate` is the interest rate per period.

`Nper` is the total number of payment periods

`Pmt` is the payment made each period; in this case, it is set to 100

`Pv` is the initial deposit; in this example, it is set to 0

`Type` is the number 0 or 1 and indicates when payments are due. Set type to 0 if payments are due at the end of the period. Set type to 1 if payments are due at the beginning. For this example, type is set to 1.

1 Make a table entering all the pertinent information as follows:

	A	B	C	D	E	F
1	Pv	Number of periods	Payment	Annual Rate	Periods per year	Periodic Rate
2	0	18	150	0.07	12	=d2/e2

2 Note that the periodic rate is calculated automatically using a formula. This gives you the flexibility to change the Number of Periods, Annual Rate and Periods per Year without having to recalculate the periodic rate.
3 Type the heading "Future Value" in cell **A4**.
4 In cell **B4**, type the formula **=fv(f2,b2,c2,a2,1)**
5 The cell references in the formula above correspond to the syntax for the Fv function. Make sure you understand each argument of the FV function and where it comes from.
6 Your finished table should look like the following.

	A	B	C	D	E	F
1	Pv	Number of Periods	Payment	Annual Rate	Periods per year	Periodic Rate
2	0	18	150	0.07	12	0.005833
3						
4	Future Value	($2,854.69)				

7 Hence, the account will be valued at $2854.69 at the end of eighteen months.

Check it out

- Change the annual rate to 10% in the above example, leaving everything else the same. What is the amount after eighteen months? after two years?

Calculating Payment for Annuities and Sinking Funds

In some cases, you will want to save a particular target amount and you are interested in how much money you should put aside each period. In Excel, this is accomplished using the PMT function. The same function can be used to calculate the amount of deposit made into a sinking fund.

Example You want to save $10000 in five years by saving a constant amount in an annuity that pays 6% interest compounded monthly. How much should you deposit in the account each month?

Solution To use the Excel function PMT, you must first know what parameters that it takes. The PMT function has the following syntax: `PMT(rate,nper,pv,fv,type)`

`Rate`	is the interest rate per period.
`Nper`	is the total number of payment periods
`Pv`	is the initial deposit; in this example, it is set to 0
`Fv`	is the future value; for this example, it is $10000
`Type`	is the number 0 or 1 and indicates when payments are due. Set type to 0 if payments are due at the end of the period. Set type to 1 if payments are due at the beginning. For this example, type is set to 0.

1 Make a table entering all the pertinent information as follows:

	A	B	C	D	E	F
1	Number of periods	Pv	Fv	Annual Rate	Periods per year	Periodic Rate
2	60	0	10000	0.06	12	`=d2/e2`

2 Note that the periodic rate is calculated automatically using a formula. This gives you the flexibility to change the Number of Periods, Annual Rate and Periods per Year without having to recalculate the periodic rate.
3 Type the heading "Monthly Payment" in cell **A4**.
4 In cell **B4**, type the formula `=pmt(f2,a2,b2,c2,0)`
5 The cell references in the formula above correspond to the syntax for the PMT function. Make sure you understand each argument of the PMT function and where it comes from.
6 Your finished table should look like the following.

	A	B	C	D	E	F
1	**Number of Periods**	**Pv**	**Fv**	**Annual Rate**	**Periods per year**	**Periodic Rate**
2	60	0	10000	0.06	12	0.005
3						
4	**Monthly Payment**	**($143.33)**				

7 The monthly payment is $143.33.

Check it out

- Recalculate the payment for the above example if the number of years is changed to seven, keeping all other values the same.

Suggested Exercises

Text	Section(s)	Exercises
Harshbarger-Reynolds *Mathematical Applications, 7th* ed.	6.3	#3-38
Berresford-Rockett *Calculus with Finite Mathematics*	1.3	#1-40, 51-60
Berresford-Rockett *Finite Mathematics*	2.3	#1-40

Present Value of Annuities

Example What lump sum would be needed on January 1 to generate annual payments of $5000 at the beginning of each year for a period of 10 years if money is worth 5.9%,compounded annually?

Solution To solve this problem, we use the PV function (for present value) in Excel. To use the Excel function PV, you must first know what parameters that it takes. The PV function has the following syntax: **`PV(rate,nper,pmt,fv,type)`**

`Rate` is the interest rate per period.

`Nper` is the total number of payment periods

`Pmt` is the payment made each period; in this case, it is set to 5000

`Fv` is the future value of the loan; in this example, it is set to 0

`Type` is the number 0 or 1 and indicates when payments are due. Set type to 0 if payments are due at the end of the period. Set type to 1 if payments are due at the beginning. For this example, type is set to 0.

1 Make a table entering all the pertinent information as follows:

	A	B	C	D	E	F
1	Number of periods	Pmt	Fv	Annual Rate	Periods per year	Periodic Rate
2	10	5000	0	0.059	1	**`=d2/e2`**

2 Note that the periodic rate is calculated automatically using a formula. This gives you the flexibility to change the Number of Periods, Annual Rate and Periods per Year without having to recalculate the periodic rate.
3 Type the heading "Present Value" in cell **`A4`**.
4 In cell **`B4`**, type the formula **`=pv(f2,a2,b2,c2,0)`**
5 The cell references in the formula above correspond to the syntax for the PV function. Make sure you understand each argument of the PV function and where it comes from.

6 Your finished table should look like the following.

	A	B	C	D	E	F
1	Number of Periods	Pmt	Fv	Annual Rate	Periods per year	Periodic Rate
2	10	5000	0	0.059	1	0.059
3						
4	Present Value	($36,975.42)				

7 The lump sum needed is $36,975.42

Suggested Exercises

Text	Section(s)	Exercises
Harshbarger-Reynolds *Mathematical Applications, 7th* ed.	6.4	#1-10, 13-32,43,44
Berresford-Rockett *Calculus with Finite Mathematics*	1.4	#1-10
Berresford-Rockett *Finite Mathematics*	2.4	#1-10

Loans and Amortization

Using Excel's payment function, one can easily calculate the payments which are due on a loan as well as calculate how much of the payment is for interest and how much for a principal.

Payment on a loan

Example A debt of $1000 with interest at 16%, compounded quarterly, is to be amortized by 20 quarterly payments (all the same size) over the next five years. What will the size of these payments be? (Example 1, Section 6.5, Harshbarger-Reynolds)

Solution We can use the same PMT function as in the previous section on annuities. We simply view it as an investment from the bank's point of view. The PMT function has the following syntax:
`PMT(rate,nper,pv,fv,type)`

`Rate` is the interest rate per period.

`Nper` is the total number of payment periods

`Pv` is the initial value of the loan; in this example, it is set to $1000

`Fv` is the future value; for this example, it is 0 since the loan will be zero at the end

`Type` is the number 0 or 1 and indicates when payments are due. Set type to 0 if payments are due at the end of the period. Set type to 1 if payments are due at the beginning. For this example, type is set to 0.

1 Make a table entering all the pertinent information as follows:

	A	B	C	D	E	F
1	Number of periods	Pv	Fv	Annual Rate	Periods per year	Periodic Rate
2	20	1000	0	0.16	4	`=d2/e2`

2 Note that the periodic rate is calculated automatically using a formula. This gives you the flexibility to change the Number of Periods, Annual Rate and Periods per Year without having to recalculate the periodic rate.

3 Type the heading "Payment" in cell **A4**.

4 In cell **B4**, type the formula `=pmt(f2,a2,b2,c2,0)`

5 The cell references in the formula above correspond to the syntax for the PMT function. Make sure you understand each argument of the PMT function and where it comes from.

6 Your finished table should look like the following.

	A	B	C	D	E	F
1	**Number of Periods**	**Pv**	**Fv**	**Annual Rate**	**Periods per year**	**Periodic Rate**
2	20	1000	0	0.16	4	0.04
3						
4	**Payment**	**($73.58)**				

7 The monthly payment is $73.58.

Check it out

- Recalculate the payment if the loan period is changed to three years.

Interest and principal payments

Example 1 A man buys a house for $200,000. He makes a $50,000 down payment and agrees to amortize the rest of the debt with quarterly payments over the next 10 years. If the interest on the debt is 12%, compounded quarterly, find

a the size of the quarterly payments,
b the size of the interest payment on the 10'th payment,
c the size of the principal payment on the 10'th payment,
d the unpaid balance immediately after the 10'th payment.

Solution

Part (a)

1 To calculate the quarterly payment, simply use the PMT function. The information is summarized below.

	A	B	C	D	E	F
1	Number of periods	Pv	Fv	Annual Rate	Periods per year	Periodic Rate
2	40	150000	0	0.12	4	`=d2/e2`

2 Type the heading "Monthly Payment" in cell **`A4`**.
3 In cell **`B4`**, type the formula **`=pmt(f2,a2,b2,c2,0)`**
4 The payment will be \$6489.36.

Part (b)

1 The size of the interest payment on the tenth payment is calculated by the IPMT function. The IPMT function has the following syntax: **`IPMT(rate,per,nper,pv,fv,type)`**

`Rate` is the interest rate per period.

`Per` is the period for which you want to find the interest and must be in the range 1 to nper

`Nper` is the total number of payment periods

`Pv` is the initial value of the loan; in this example, it is set to \$150000

`Fv` is the future value; for this example, it is 0 since the loan will be zero at the end

`Type` is the number 0 or 1 and indicates when payments are due. Set type to 0 if payments are due at the end of the period. Set type to 1 if payments are due at the beginning. For this example, type is set to 0.

2 Type the heading "Interest Payment- period 10" in cell **`A6`**.
3 In cell **`B6`**, type the formula **`=ipmt(f2,10,a2,b2,c2,0)`**
4 The interest payment will be \$3893.70.

Part (c)

1 The size of the principal payment on the tenth payment is calculated by the PPMT function. The PPMT function has the following syntax: **`PPMT(rate,per,nper,pv,fv,type)`**
2 The definitions of the parameters are the same as for the IPMT function above and will not be repeated here.
3 Type the heading "Principal Payment- period 10" in cell **`A8`**.
4 In cell **`B8`**, type the formula **`=ppmt(f2,10,a2,b2,c2,0)`**
5 The principal payment will be \$2595.66.

Part (d)

1 The unpaid balance is simply the present value of an annuity consisting of 30 payments. Hence we use the PV function in Excel.
2 To use the Excel function PV, you must first know what parameters that it takes. The PV function has the following syntax: **`PV(rate,nper,pmt,fv,type)`**

`Rate`	is the interest rate per period.
`Nper`	is the total number of payment periods - in this case 30
`Pmt`	is the payment made each period; in this case, it is set to the cell reference **`b4`**
`Fv`	is the future value of the loan; in this example, it is set to 0
`Type`	is the number 0 or 1 and indicates when payments are due. Set type to 0 if payments are due at the end of the period. Set type to 1 if payments are due at the beginning. For this example, type is set to 0.

3 Type the heading "Unpaid balance- period 10" in cell **`A10`**.
4 In cell **`B10`**, type the formula **`=pv(f2,30,b4,0,0)`**
5 The unpaid balance will be $127,194.26

Check it out

- Repeat the example above if the loan period is changed to 15 years, leaving all other parameters the same.

Suggested Exercises

Text	Section(s)	Exercises
Harshbarger-Reynolds *Mathematical Applications, 7th* ed.	6.5	#3-26, 31-38
Berresford-Rockett *Calculus with Finite Mathematics*	1.4	#11-30, 41-60
Berresford-Rockett *Finite Mathematics*	2.4	#11-38

Chapter 8

Probability and Statistics

Calculating quantities using the binomial and normal distributions is easily accomplished with Excel. Also, the charting features and the built-in statistical features of Excel make the analysis of large sets of data more tractable. The following sections will illustrate how to use Excel in a variety of topics in probability and statistics.

Binomial Probability

Example 1 A die is rolled 4 times and the number of times a 6 results is recorded. What is the probability that three 6's will result? (Example 1, Section 8.1, Harshbarger-Reynolds.)

Solution For this experiment, a success is rolling a 6. The probability of success is 1/6 and the probability of not rolling a 6, i.e. a failure, is 1-1/6=5/6. There are 4 trials of this experiment. We are interested in the probability of exactly 3 successes.

Steps to calculating binomial probabilities

1 Type headings in cells **`a1:a3`** and their respective values in cells **`b1:b3`** as follows:

	A	B
1	Number of successes	3
2	Number of trials	4
3	Probability of success	0.16666667

2 To calculate the probability of 3 successes, we use the formula for binomial probabilities. To do this in Excel, we use the built-in Excel function **`binomdist`**. The syntax for binomdist is as follows: **`binomdist(number_s,trials,probability_s,cumulative)`**

- a **`number_s`** is the number of successes in trials.
- b **`trials`** is the number of independent trials.
- c **`probability_s`** is the probability of success on each trial.
- d **`cumulative`** is set to **`false`** if you are interested only in the probability of exactly **`number_s`** successes. It is set to **`true`** if you are interested in less than or equal to **`number_s`** successes.

3 In cell **`A4`**, type the heading "Probability of 3 successes".

4 In cell **`B4`**, type the formula **`=binomdist(b1,b2,b3,false)`**

	A	B
4	Probability of 3 successes	=binomdist(b1,b2,b3,false)

5 Explanation:

a **b1** contains the value for the number of successes
b **b2** contains the value for the number of trials
c **b3** contains the value for the probability of success
d the last argument is set to **false** since we want the probability of *exactly* three successes.

6 The calculated probability will be 0.015432.

Check it out

- Calculate the probability of rolling exactly 2 sixes (Answer: 0.115741)
- Calculate the probability of rolling at most 2 sixes. *Hint:* set the **cumulative** value to **true**. (Answer: 0.9838)

Example 2 A manufacturer of motorcycle parts guarantees that a box of 24 parts will contain at most 1 defective part. If the records show that the manufacturer's machines produce 1% defective parts, what is the probability that a box of parts will satisfy the guarantee? (Example 3, Section 8.1, Harshbarger-Reynolds)

Solution For this experiment, a success is getting a defective part. The probability of success is 0.01. There are 24 trials of this experiment. We are interested in the probability of *at most* 1 success.

Steps to calculate binomial probabilities

1 Type headings in cells **a1:a3** and their respective values in cells **b1:b3** as follows:

	A	B
1	Number of successes	1
2	Number of trials	24
3	Probability of success	0.01

2 In cell **A4**, type the heading "Probability of at most 1 success".

3 We use the **binomdist** function, described in detail in the first example. But now, since we want at most 1 success, the **cumulative** option is set to **true**.

4 In cell **B4**, type the formula **=binomdist(b1,b2,b3,true)**

	A	B
4	Probability of at most 1 success	=binomdist(b1,b2,b3,true)

5 The calculated probability will be 0.9762.

Suggested Exercises

Text	Section(s)	Exercises
Harshbarger-Reynolds *Mathematical Applications, 7th* ed.	8.1	#15-33
Berresford-Rockett *Calculus with Finite Mathematics*	4.4	#25-27;#37-44
Berresford-Rockett *Finite Mathematics*	5.6	#25-31;#38-44

Descriptive Statistics

Frequency Tables and Histograms

Excel's charting capabilities can be used to generate frequency histograms for sets of data.

Example Construct a histogram for the following breakdown of test scores.

Grade Range	Frequency
90-100	2
80-89	5
70-79	7
60-69	3
0-59	2

Solution A histogram is created using Chart Wizard in a manner similar to creating graphs of functions.

Steps to creating a histogram

1 Copy the entries of the table above to cells `a1:b6`.
2 Select the range `a1:b6`.
3 Click on the Chart Wizard icon.

4 Select the column graph option with the first sub-type.

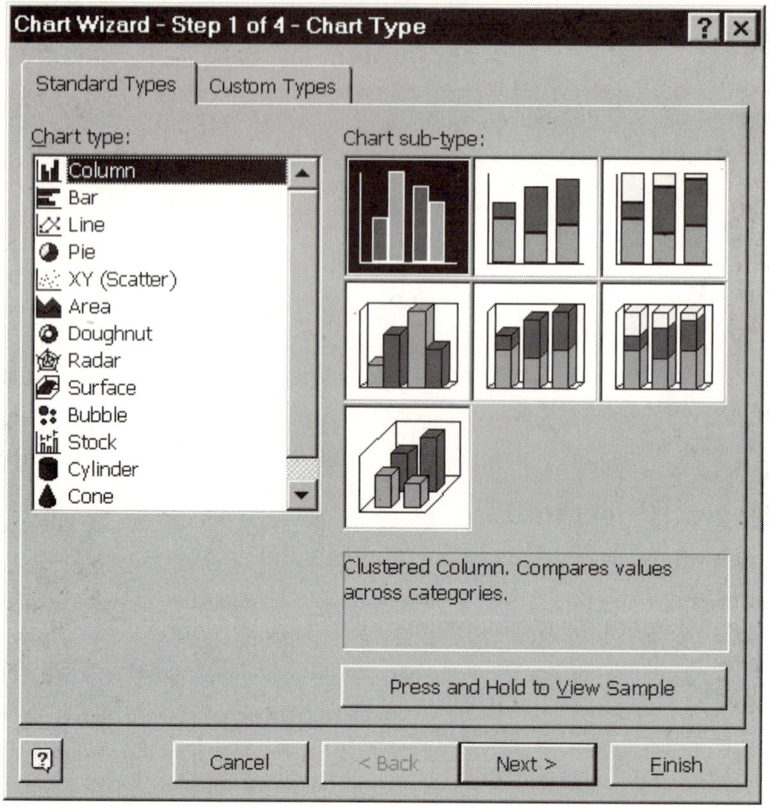

5 Click Next.
6 Click Next through Steps 2-4. You may want to add a title etc. in Step 3.
7 Your histogram will look like the following.

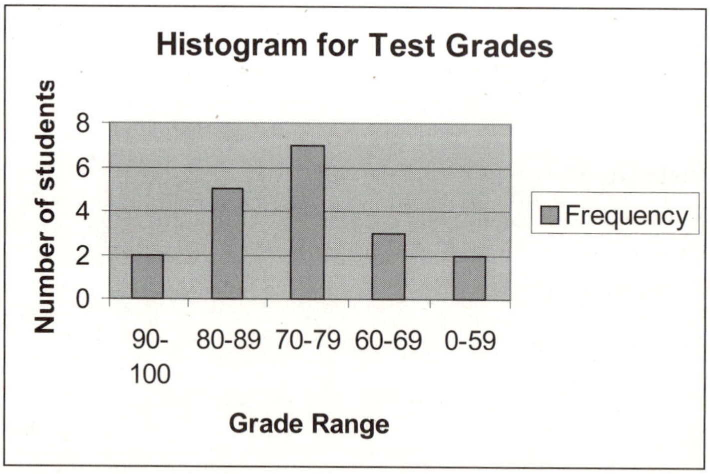

Check it out

- Change some the numbers in the frequency column in your worksheet. What happens to your chart?

Suggested Exercises

Text	Section(s)	Exercises
Harshbarger-Reynolds *Mathematical Applications, 7th* ed	8.2	#1-6
Berresford-Rockett *Calculus with Finite Mathematics*	5.1	#21-30
Berresford-Rockett *Finite Mathematics*	6.1	#21-30

Finding the mean and standard deviation using Excel tables

Example Find the mean and standard deviation of the following sample of test scores

Grade Range	Class Marks	Frequency
90-100	95	3
80-89	84.5	4
70-79	74.5	7
60-69	64.5	0
50-59	54.5	2

Solution Note that in finding the mean when data sets are given in intervals, we use the class mark (midpoint) to represent the data in the interval.

Steps to finding the mean

1 Enter the data above in cells **`a1:c6`**.
2 In **`d1`**, type the heading "Class mark * frequency".
3 In cell **`d2`**, type the formula **`=b2*c2`**
4 Copy the formula in **`d2`** to **`d3:d6`**
5 In cell **`b7`**, type the heading "Total"
6 In cell **`c7`**, type the formula for the total frequencies **`=sum(c2:c6)`**
7 The Excel function sum simply adds up the values in the given range of cells. Your table should look like the following (in formula view).

	A	B	C	D
1	Grade Range	Class marks	frequency	class mark*frequency
2	90-100	95	3	=B2*C2
3	80-89	84.5	4	=B3*C3
4	70-79	74.5	7	=B4*C4
5	60-69	64.5	0	=B5*C5
6	50-59	54.5	2	=B6*C6
7		Total	=SUM(C2:C6)	
8				

8 In cell **`d7`**, type the formula for the total of "Class mark * frequency" **`=sum(d2:d6)`**

9 In cell **a8**, type the heading "Mean"
10 In cell **a9**, type the formula **=d7/c7**
11 Explanation: cell **a9** calculates the average by taking the total of "class mark * frequency" and dividing it by the total frequency.
12 The completed table will look like the following.

	A	B	C	D	E
1	Grade Range	Class marks	frequency	class mark*frequency	
2	90-100	95	3	285	
3	80-89	84.5	4	338	
4	70-79	74.5	7	521.5	
5	60-69	64.5	0	0	
6	50-59	54.5	2	109	
7		Total	16	1253.5	
8	Mean				
9	78.344				

Steps to finding the standard deviation

1 To calculate the standard deviation, you will first need to subtract each data point from the mean and square it. In cell **e1**, type the heading "freq*(x-x_mean)^2".
2 In cell **e2**, type the formula **=c2*(b2-a9)^2**
3 Explanation: since we do not want the cell reference for the mean to change as we copy down the column, its *absolute reference* is given by **a9**, rather than **a9**.
4 Copy the formula in **e2** to **e3:e6.**
5 In cell **e7**, type the formula for the total of "freq*(x-x_mean)^2" **=sum(e2:e6)**
6 In cell **a10**, type the heading "Standard Deviation"
7 In cell **a11**, type the formula **=sqrt(e7/(c7-1))**
8 Note that we divide by n-1=total frequencies - 1 in the above step. The formulas are shown in the table below.

	A	B	C	D	E
1	Grade Range	Class marks	frequency	class mark*frequency	freq*(x-x_mean)^2
2	90-100	95	3	=B2*C2	=C2*(B2-$A$9)^2
3	80-89	84.5	4	=B3*C3	=C3*(B3-$A$9)^2
4	70-79	74.5	7	=B4*C4	=C4*(B4-$A$9)^2
5	60-69	64.5	0	=B5*C5	=C5*(B5-$A$9)^2
6	50-59	54.5	2	=B6*C6	=C6*(B6-$A$9)^2
7		Total	=SUM(C2:C6)	=SUM(D2:D6)	=SUM(E2:E6)
8	Mean				
9	=D7/C7				
10	Standard Deviation				
11	=SQRT(E7/(C7-1))				
12					

9 The completed table will look like the following

	A	B	C	D	E
1	Grade Range	Class marks	frequency	class mark*frequency	freq*(x-x_mean)^2
2	90-100	95	3	285	832.2919922
3	80-89	84.5	4	338	151.5976563
4	70-79	74.5	7	521.5	103.4208984
5	60-69	64.5	0	0	0
6	50-59	54.5	2	109	1137.048828
7		Total	16	1253.5	2224.359375
8	Mean				
9	78.344				
10	Standard Deviation				
11	12.177				

10 Hence, we see that the class average is 78.344 and the standard deviation is 12.177.

Check it out

- Change the frequency to 4,3,3,2,4, from top to bottom. What happens to the standard deviation? Why?

Excel functions for statistics

Given a raw data set, Excel has built-in functions to find the mean, median, and standard deviation. The following example illustrates the use of these functions.

Example The following table gives the average hourly earnings of production workers in private industry from January 1999 to January 2000. (Source: Bureau of Labor Statistics). Calculate the mean hourly earnings over the thirteen months as well as the standard deviation. Also calculate the median.

Table 1 Average Hourly Earnings

	Jan	Feb	Mar	Apr	May	Jun	Jul	Aug	Sep	Oct	Nov	Dec
1999	13.11	13.10	13.12	13.16	13.19	13.14	13.15	13.20	13.38	13.41	13.43	13.47
2000	13.58											

Solution The Excel functions for the mean, standard deviation and median are **`average`**, **`stdev`**, and **`median`**, respectively. They take as their arguments the ranges of cell references containing the data.

1 Enter the data above in cells **`a1:m1`** as shown below

	A	B	C	D	E	F	G	H	I	J	K	L	M
1	13.11	13.1	13.12	13.16	13.19	13.14	13.15	13.2	13.38	13.41	13.43	13.47	13.58
2													

2 In cell **`a3`**, type the heading “Average”

3 In cell **`b3`**, type the formula **`=average(a1:m1)`**

4 In cell **a4**, type the heading "Standard deviation"
5 In cell **b4**, type the formula **=stdev(a1:m1)**
6 In cell **a5**, type the heading "Median"
7 In cell **b5**, type the formula **=median(a1:m1)**
8 Your finished table will resemble the following

	A	B	C	D	E	F	G	H	I	J	K	L	M
1	13.11	13.1	13.12	13.16	13.19	13.14	13.15	13.2	13.38	13.41	13.43	13.47	13.58
2													
3	Average	13.26462											
4	Standard deviation	0.1645											
5	Median	13.19											

9 Note that the average and the median are not the same. The average is slightly higher than the median since the higher earnings toward the end of the data range push up the average.

Suggested Exercises

Text	Section(s)	Exercises
Harshbarger-Reynolds *Mathematical Applications, 7th* ed.	8.2	#49-54
Berresford-Rockett *Calculus with Finite Mathematics*	5.2 5.3	#1-20 #1-25
Berresford-Rockett *Finite Mathematics*	6.2 6.3	#1-20 #1-25

Normal Probabilities

Example If IQ scores follow a normal distribution with mean 100 and standard deviation 15, what percentage of the scores will be between 100 and 115? (Example 1(a), Section 8.5, Harshbarger-Reynolds.)

Solution Excel has a built-in function **normdist**, which will calculate the above probability.

Steps to calculate normal probabilities

1 Type headings in cells **a1:a4** and their respective values in cells **b1:b4** as follows:

	A	B
1	Mean	100
2	Standard deviation	15
3	x1	100

	A	B
4	x2	115

2 To calculate Pr(x1 < X < x2), we first calculate Pr(X < x1) and subtract it from Pr(X < x2). To do this in Excel, we use the built-in Excel function **`normdist`**. The syntax for normdist is as follows: **`normdist(a,mean,standard deviation,cumulative)`**

- **a** **`a`** is the number such that Pr(X < a) is calculated.
- **b** **`mean and standard deviation`** are parameters of the given normal distribution.
- **c** **`cumulative`** is set to **`true`** if you are interested in the normal probability less than or equal to **a**.

3 In cells **A5:B7**, type the following headings and formulas:

	A	B
5	Pr(X < x1)	**`=normdist(b3,b1,b2,true)`**
6	Pr(X < x2)	**`=normdist(b4,b1,b2,true)`**
7	Pr(x1 <X <x2)	**`=b6-b5`**

4 Explanation:

- **a** **`b1`** contains the value for the mean
- **b** **b2** contains the value for the standard deviation
- **c** **b3** contains the value for **a** such that Pr(X < **a**) is calculated; similarly for **b4**
- **d** the last argument is set to true since we are interested in the normal probability less than or equal to **a**.
- **e** In **b7**, we simply type the formula to subtract the two probabilities.

5 The results are as follows:

	A	B
1	**Mean**	100
2	**Standard Deviation**	15
3	**x1**	100
4	**x2**	115
5	**Pr(X < x1)**	0.5
6	**Pr(X < x2)**	0.841345
7	**Pr(x1<X<x2)**	0.341345
8		

6 The calculated probability will be 0.341345. This means that roughly 34% of the scores will be between 100 and 115.

Check it out

- What percentage of IQ scores will be between 80 and 100? (Answer: 41%)
- What percentage of IQ scores will be *greater* than 120? (Answer: 9.1%)

Suggested Exercises

Text	Section(s)	Exercises
Harshbarger-Reynolds *Mathematical Applications, 7th* ed.	8.5	#27-38
Berresford-Rockett *Calculus with Finite Mathematics*	5.4	#1-40
Berresford-Rockett *Finite Mathematics*	6.4	#1-40

Chapter 9

Limits and Derivatives

Numerical Investigation of Limits

We can make of tables of function values with Excel to easily investigate the behavior of a function near a point. Review the section on making tables in Excel in Chapter 1, if necessary.

Example 1 Find the limit, if it exists, of $f(x)=(x^2-4)/(x-2)$ at x=2.

Solution The function is not defined at x=2. Therefore, make a table of function values near x=2.

1 Limit from the right: first, let x approach 2 from the right.
2 Use x= 2.1,2.05,2.01,2.001,2.0001. You will get the table below.

	A	B
1	x	f(x)
2	2.1	4.1
3	2.05	4.05
4	2.01	4.01
5	2.001	4.001
6	2.0001	4.0001

3 Limit from the left: next, let x approach 2 from the left. Use x=1.9,1.95,1.99,1.999,1.9999. You will get the table below.

C	D
x	f(x)
1.9	3.9
1.95	3.95
1.99	3.99
1.999	3.999
1.9999	3.9999

4 From the numerical results above, we can conclude that the limit is 4.

Check it out

- Find the limit, if it exists, of $f(x)=(x^2-16)/(x+4)$ at x=-4.

Example 2 Find the limit, if it exists, of $f(x)=|x|/x$ at x=0.

Solution The function is not defined at x=0. Therefore, make a table of function values near x=0. Use the built-in Excel function **`ABS`** for absolute value: to calculate the function value of the x-value in cell **`A1`**, in cell **`B1`**, type **`=abs(a1)/a1`**

Copy the formula down the column for the other function values.

1 Limit from the right: first, let x approach 0 from the right.
2 Use x= 0.1,0.05,0.01,0.001,0.0001. You will get the table below.

	A	B
1	x	f(x)
2	0.1	1
3	0.05	1
4	0.01	1
5	0.001	1
6	0.0001	1

3 Limit from the left: next, let x approach 2 from the left.
4 Use x=-0.1,-0.05,-0.01,-0.001,-0.0001. You will get the table below.

C	D
x	f(x)
-0.1	-1
-0.05	-1
-0.01	-1
-0.001	-1
-0.0001	-1

5 From the numerical results above, we see that the limit from the right is 1 and the limit from the left is -1. Since the two values are not equal, the limit of f(x)=|x|/x at zero does not exist.

Check it out

- Graph the function in the above example near x=0 to see that the limit does not exist.
- Investigate the limit as x approaches zero of $f(x)=(x^2-x)/x$.

Suggested Exercises

Text	Section(s)	Exercises
Harshbarger-Reynolds *Mathematical Applications,* 7th ed.	9.1	#13-16, #39-51
Larson-Edwards *Calculus: An Applied Approach,* 6th ed. (and brief version)	1.5	#1-8,55-67
Berresford-Rockett *Applied Calculus,* 3rd ed. (and brief version)	2.1	#3-12,79,80
Berresford-Rockett *Calculus with Finite Mathematics*	6.1	#3-12,77,78

Numerical Investigation of Derivatives

Recall that the definition of a derivative of f at a point x is

$$\lim_{h \to 0} \frac{f(x+h) - f(x)}{h} = f'(x)$$

In Excel, you can investigate this limit numerically to see what the derivative at a point x will be. The quantity $\frac{f(x+h) - f(x)}{h}$ is called a *difference quotient*. These are simply the values for the slopes of the corresponding secant lines.

Example Investigate the derivative of $f(x) = x^3$ at x=1 as the limit of the difference quotients.

Solution

1 Set up the first row of a table of values for calculating the difference quotients. Enter the formulas as shown.

	A	B	C	D	E
1	h	1+h	f(1)	f(1+h)	(f(1+h)-f(1))/h
2	0.1	=1+a2	=1^3	=b2^3	=(d2-c2)/a2

2 Next, type values of h tending to 0 down column A. Values of h must be both positive and negative. Copy the formulas is **b2:e2** downwards. A finished table shown with all the formulas is as follows:

	A	B	C	D	E
1	h	1+h	f(1)	f(1+h)	(f(1+h)-f(1))/h
2	0.1	=1+A2	=1^3	=B2^3	=(D2-C2)/A2
3	0.01	=1+A3	=1^3	=B3^3	=(D3-C3)/A3
4	0.001	=1+A4	=1^3	=B4^3	=(D4-C4)/A4
5	0.0001	=1+A5	=1^3	=B5^3	=(D5-C5)/A5
6	-0.0001	=1+A6	=1^3	=B6^3	=(D6-C6)/A6
7	-0.001	=1+A7	=1^3	=B7^3	=(D7-C7)/A7
8	-0.01	=1+A8	=1^3	=B8^3	=(D8-C8)/A8
9	-0.1	=1+A9	=1^3	=B9^3	=(D9-C9)/A9
10					

3 The values are as follows:

	A	B	C	D	E
1	h	1+h	f(1)	f(1+h)	(f(1+h)-f(1))/h
2	0.1	1.1	1	1.331	3.31
3	0.01	1.01	1	1.030301	3.0301
4	0.001	1.001	1	1.003003	3.003001
5	0.0001	1.0001	1	1.0003	3.00030001
6	-0.0001	0.9999	1	0.9997	2.99970001
7	-0.001	0.999	1	0.997003	2.997001
8	-0.01	0.99	1	0.970299	2.9701
9	-0.1	0.9	1	0.729	2.71

4 From the numbers in the table above, a h approaches zero, the values of the difference quotients approach 3 *from both positive and negative values of h.* Thus we can conclude that the values of f'(1) is 3. The actual value of f'(1) is $3(1)^2=3$, evaluated using the power rule.

Check it out

- Numerically investigate the value of the derivative of $f(x) = 3x^2-1$ at x=2. Remember to change all your formulas accordingly if you are modifying the table used for the above example. Compare your result with the exact answer for this problem.
- Decide if f(x)=|x| is differentiable at x=0 by examining the limits of the difference quotients.

Suggested Exercises

Text	Section(s)	Exercises
Harshbarger-Reynolds *Mathematical Applications,* 6th ed.	9.3	#13-18,#41-53
Larson-Edwards *Calculus: An Applied Approach,* 6th ed. (and brief version)	2.1	#27-36 (numerically investigate)
Berresford-Rockett *Applied Calculus,* 3rd ed. (and brief version)	2.2	#17-20 (numerically investigate)
Berresford-Rockett *Calculus with Finite Mathematics*	6.2	#39,40

Derivatives and Rates of Change

Using the ideas in the section above on the numerical investigation of derivatives, we can calculate such quantities as average velocity and marginals.

Example If a free-falling object is dropped from a height of 100 feet, and air resistance is neglected, the height h (in feet) of the object at time t (in seconds) is given by

$$h(t) = -16t^2 + 100.$$

Find the average velocity of the object in the interval [1,1.5]. (Example 2, Section 2.3, Larson-Edwards)

Solution

1 Set up a table with formulas as shown below

	A	B	C	D	E
1	t1	t2	f(t1)	f(t2)	(f(t2)-f(t1))/)(t2-t1)
2	1	1.5	=-16*a2^2+100	=-16*b2^2+100	=(d2-c2)/(b2-a2)

2 Your finished table will look as follows

	A	B	C	D	E
1	**t1**	**t2**	**f(t1)**	**f(t2)**	**(f(t2)-f(t1))/(t2-t1)**
2	1	1.5	84	64	-40

3 From the above table, we see that the average velocity over the interval [1,1.5] is -40 ft./sec.

Check it out

- Extend the table above to calculate the average velocity over the following intervals:

a [1,1.2]
b [1,1.1]
c [1,1.05]
d [1,1.01]

- What do you observe about the values of the average velocity as the intervals get smaller?

Suggested Exercises

Text	Section(s)	Exercises
Harshbarger-Reynolds *Mathematical Applications,* 7th ed.	9.3	#39-42
Larson-Edwards *Calculus: An Applied Approach,* 6th ed. (and brief version)	2.3	#9-10,34-38
Berresford-Rockett *Applied Calculus,* 3rd ed. (and brief version)	2.2	#9-16

Chapter 10

Graphs of Functions and their Derivatives

Using Excel to make tables and graphs of functions and their derivatives can help you understand their relationships. Review Chapters 1 and 2 on making tables and graphs before proceeding with this section.

A Function and its First Derivative

Example Graph the function $f(x)=x^2$ and its derivative $f'(x)=2x$ on the interval [-2,2] in one plot. Use the graph to see where the function is increasing and decreasing.

Solution

1 Make a table with a row for headings and a second row with formulas as follows:

	A	B	C
1	x	f(x)	f'(x)
2	-2	=a2^2	=2*a2

2 Extend the table for values of x from -2 to 2 in increments of 0.5. Refer to instructions on making tables in Chapter 1.

3 Extend formulas for f(x) and f'(x) down the column. Your finished table should be similar to the following.

	A	B	C
1	x	f(x)	f'(x)
2	-2	4	-4
3	-1.5	2.25	-3
4	-1	1	-2
5	-0.5	0.25	-1
6	0	0	0
7	0.5	0.25	1
8	1	1	2
9	1.5	2.25	3
10	2	4	4

4 Select the entire table and graph. Follow the directions for graphing multiple functions in Chapter 2.

5 Your graph will look like the following.

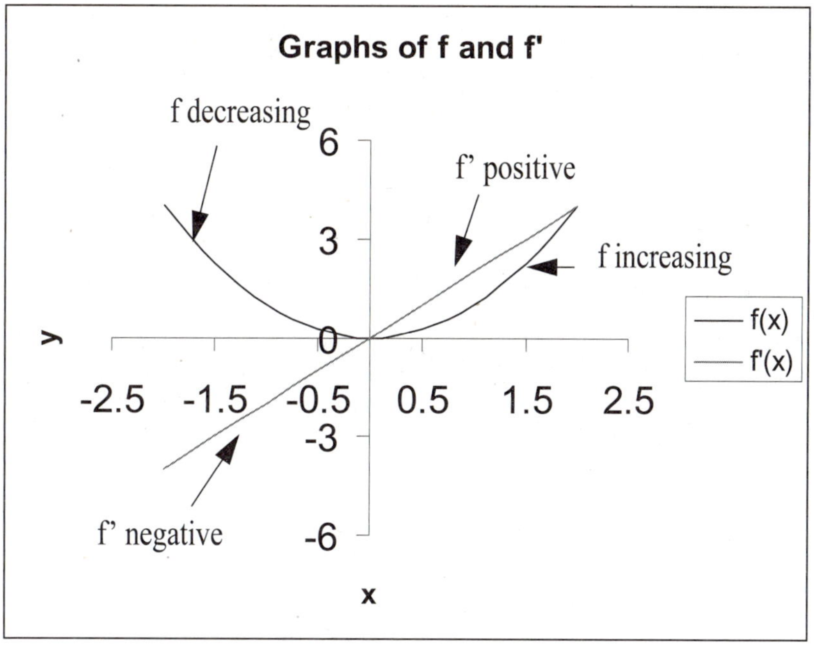

6 From the graph, we see that on (-2,0), f(x) is decreasing. On this interval, the values of f'(x) are negative.

7 f(x) is increasing on (0,2) since the values of f'(x) on this interval are positive.

8 The sign of the derivative changes at x=0, where f'(0)=0.

Check it out

- Perform a similar analysis as in the above example with the function $f(x)=x^3-x$ on the interval [-2,2]. You may use Goal Seek to locate the zeros of f'(x).

Suggested Exercises

Text	Section(s)	Exercises
Harshbarger-Reynolds *Mathematical Applications,* 7th ed.	10.1	#37-42,62-65
Larson-Edwards *Calculus: An Applied Approach,* 6th ed. (and brief version)	3.1 3.2	#19-28,35,39 # 13-18,27-34,43,45
Berresford-Rockett *Applied Calculus,* 3rd ed. (and brief version)	3.1	#29-40
Berresford-Rockett *Calculus with Finite Mathematics*	7.1	#29-40

A Function and its Second Derivative

Example Graph f, f', and f" on the same plot for $f(x) = x^3-9x^2+24x$ on the interval [0,5]. Discuss where the function has relative extrema and where the concavity changes. (Example 4, Sec. 3.2, Berresford-Rockett)

Solution

1 For this function, $f'(x)=3x^2-18x+24$ and $f''(x)=6x-18$.
2 Make a table with a row for headings and a second row with formulas as follows:

	A	B	C	D
1	x	f(x)	f'(x)	f"(x)
2	0	=a2^3-9*a2^2+24*a2	=3*a2^2-18*a2+24	=6*a2-18

3 Extend the table for values of x from 0 to 5 in increments of 0.5. Also extend the formulas for f, f' and f" down the columns. Refer to instructions on making tables in Chapter 1.
4 Your finished table should be similar to the following.

	A	B	C	D
1	x	f(x)	f'(x)	f"(x)
2	0	0	24	-18
3	0.5	9.875	15.75	-15
4	1	16	9	-12
5	1.5	19.125	3.75	-9
6	2	20	0	-6
7	2.5	19.375	-2.25	-3
8	3	18	-3	0
9	3.5	16.625	-2.25	3
10	4	16	0	6
11	4.5	16.875	3.75	9
12	5	20	9	12

5 Select the entire table and graph. Follow the directions for graphing multiple functions in Chapter 2.

6 Your graph will be similar to the following.

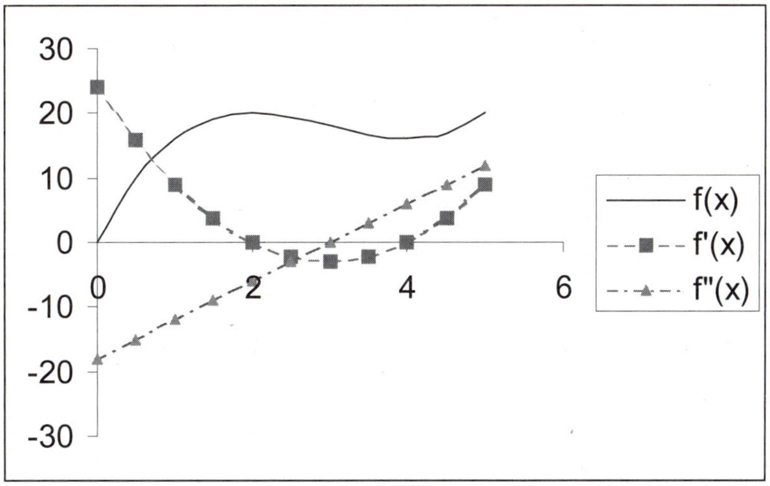

7 Reading from the graph, f'(2)=f'(4)=0. These are candidates for relative extrema. Since f"(2) < 0, f has a relative maximum at x=2. Since f"(4) > 0, f has a relative minimum at x=4.

8 f is increasing on (0,2), decreasing on (2,4) and increasing on (4,5). This is seen by the values of f' on these intervals, which are positive, negative, and positive, respectively.

9 f" is negative on (0,3) and so f is concave down on (0,3). f" is positive on (3,5) and so f is concave up on (3,5). Since f" changes sign at x=3, x=3 is an inflection point. This translates to a change of concavity for f at x=3.

Check it out

- Graph and perform an analysis similar to the example above for the function

  $f(x) = x^3 - 4x$ on the interval [-3,3].

Suggested Exercises

Text	Section(s)	Exercises
Harshbarger-Reynolds *Mathematical Applications,* 7th ed.	10.2	#25,26,41-45
Larson-Edwards *Calculus: An Applied Approach,* 6th ed. (and brief version)	3.3	#19-22, 35-46,51-54,57,58,60-65
Berresford-Rockett *Applied Calculus,* 3rd ed. (and brief version)	3.2	#27-36
Berresford-Rockett *Calculus with Finite Mathematics*	7.2	#27-36

Chapter 11

Optimization in One Variable Using Solver

This chapter will illustrate the use of an Excel tool called Solver to solve optimization problems from calculus. To check that your installation of Excel has Solver, select the Tools menu bar. If you see Solver as one of the options, you are ready to go. Otherwise, you will have to add it in. See the preface on how to add in Solver.

Example A packaging company wishes to design an open top box with a square base whose volume is exactly 40 cubic feet. Find the dimensions of the box requiring the least amount of materials.

Solution The problem must first be set up by hand before we can use Excel Solver.

Set up of problem

1 Identify variables

x: length of side of base

y: height of box

2 Identify objective: Minimize surface area. Surface area = (area of base) + (area of four sides) = $A = x^2 + 4xy$

3 We must express the area as a function of only one variable. To do this, use the information about the volume: $V = x^2 y = 40$

4 Solving for y in terms of x in the above equation gives $y = \frac{40}{x^2}$.

5 Now, we can write A in terms of x. Plug in the expression for y in Step 4 into the area formula in Step 2 to get

$$A = x^2 + 4x\left(\frac{40}{x^2}\right) = x^2 + \frac{160}{x}$$

6 This will be the expression we wish to minimize.

Graph the function being minimized

1 It is a good idea to graph the function which you are trying to minimize so that you have an idea of where a minimum would occur.

2 Note that deciding the interval on which to graph the function requires some thinking. Looking at the area function, we can see that making x too small will make the area large and also making x large will make the area large. The following table lists some values for the area in [1,10]. The area is first decreasing and then increasing. From the table, it looks like the minimum lies somewhere between 3 and 5

x	1	2	3	4	5	6	7	8	9	10
A	161	84	62.3	56	57	62.7	71.9	84	98.8	116

3 Graph this function on [1,10] using the directions in Chapter 2. Note that the function is not defined at x=0. You should obtain a graph resembling the following:

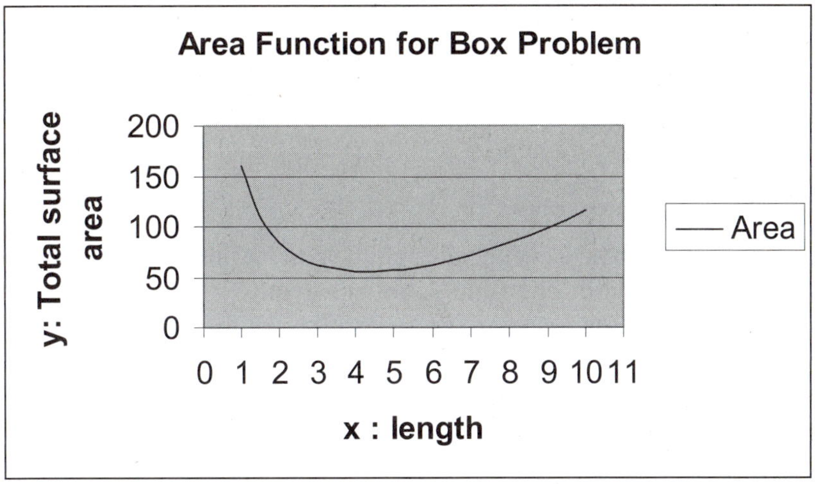

4 The graph also shows that a minimum is somewhere between x=3 and x=5.

The next step is to input all this information into Excel so that Solver can be invoked. Since all the calculations in the spreadsheet are done with cell references, you must set up cell entries for the variable and the function you are optimizing. In Excel, the cell containing the formula for the optimizing function is referred to as the *target cell*. The cells containing the variables are called the *changing cells*.

Steps to set up the problem in Excel

1 In a blank spreadsheet, first type a heading called "Variable" in cell **a1**, followed by the variable description in **a3** and its value in cell **b3**. The variable x is initially assigned a value of 1. Refer to the table below as a guide.

	A	B
1	Variable	
2		
3	x: length of base	1

2 The area function formula is given in terms of the cell references for the variables x. Enter the information for the area function as follows:

a Type a heading called "Objective" in cell **a5**
b Type a description of the objective in cell **a7**
c Enter the objective function formula in **b7** .The formula is **=b3^2+160/b3** .

3 Your completed table should look like the one below (shown using formula view).

	A	B
1	**Variable**	
2		
3	x:length of base	1
4		
5	**Objective**	
6		
7	Minimize Area	=B3^2+160/B3

Check it out

- To get familiar with the setup of the problem in Excel, change the value of the x variable in **`b3`** to some other values. What happens to the value of the objective function?

Steps to solve the problem using Solver

1 Once you check that your spreadsheet contains all the correct formulas in the appropriate cells, you are ready to invoke Solver by choosing **Tools > Solver**.

2 You will see a dialog box whose first entry is the information for the target cell (i.e. the objective function). Click cursor into the this entry box and click into cell **`B7`** (formula for objective function).

3 Check the button to minimize. You dialog box should now look like the one below.

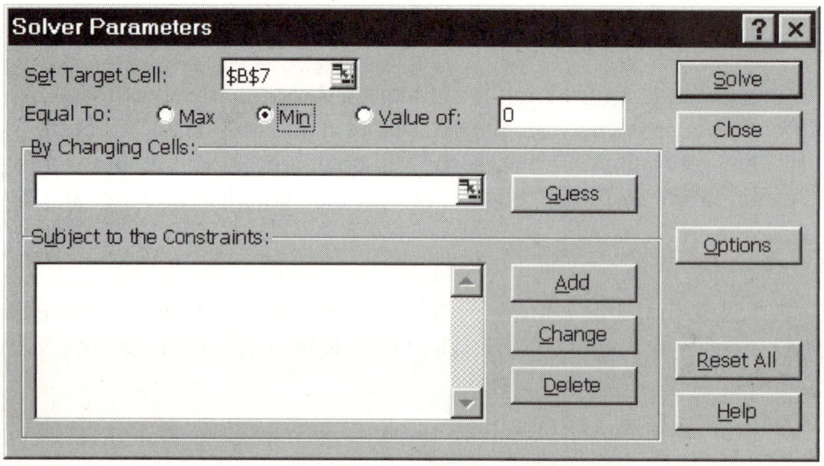

4 Next click cursor to the By Changing Cells entry box.

5 Enter the cell reference for the x variable by selecting the cell **`b3`** . This portion of your dialog box should resemble the following.

6 Your completed Solver box should resemble the following.

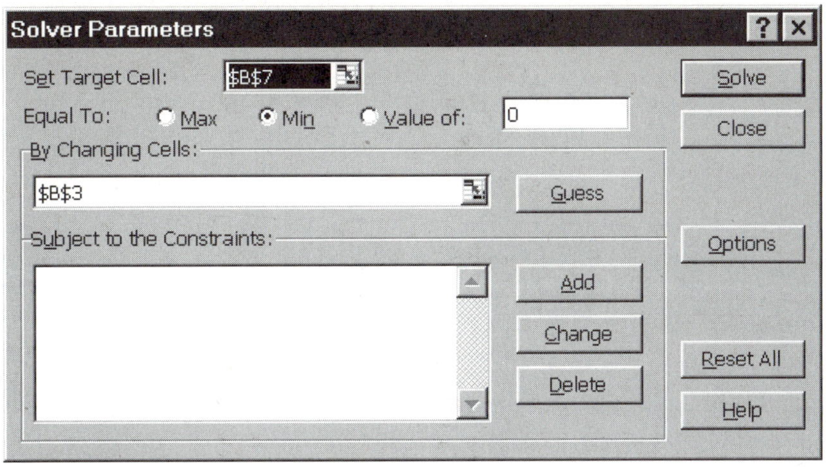

7 Click into the Options box, and make sure that the Assume Linear Model checkbox is NOT checked, since this problem is NOT linear. See the following figure.

should NOT be checked

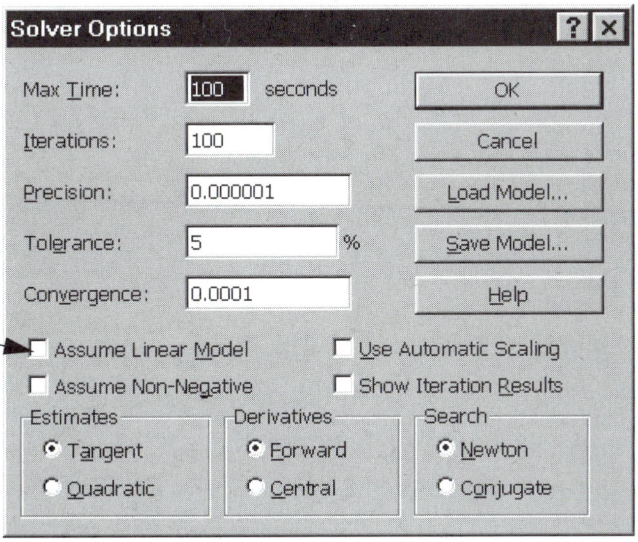

8 Click OK in the Options dialog box.
9 Click Solve in the Solver dialog box. You will get a dialog box stating that Solver found a solution.
10 Check the Keep Solver Solution button and also select the Answer report. Your completed box resembles the following.

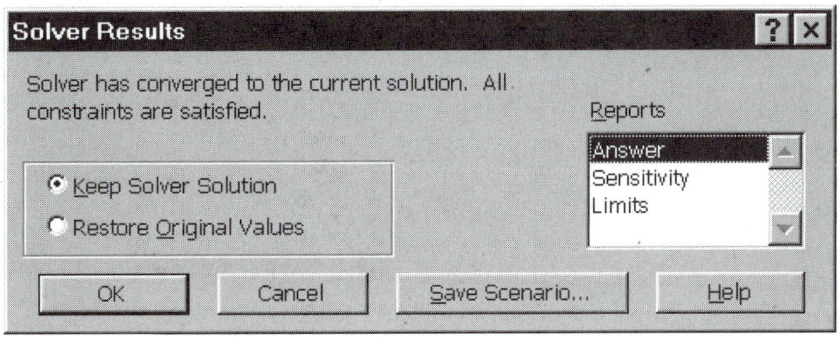

11 Click OK. Go back and examine the cells with the variable and objective. They should now contain the optimal values and resemble the following table.

	A	B
1	**Variable**	
2		
3	x:length of base	4.3089
4		
5	**Objective**	
6		
7	Minimize Area	55.699

12 The area is minimized for x=4.31 and the minimal area is 55.699. The height of the box, y, is given by $40/x^2 = 40/(4.31)^2 = 2.15$.

13 As we had observed from the table and graph of the area function, the x value producing the minimum area is indeed between 3 and 5.

14 Click on the worksheet labeled Answer Report 1 to see a summary of the solution.

Suggested Exercises

Text	Section(s)	Exercises
Harshbarger-Reynolds	10.3	#5-20, 27-30,33-45
Mathematical Applications, 7th ed.	10.4	#11-21, 23-32
Larson-Edwards	3.4	#13-37
Calculus: An Applied Approach, 6th ed.	3.5	6th ed.#13-40
(and brief version)		
Berresford-Rockett	3.3	#21-37, 43-50
Applied Calculus, 3rd ed.	3.4	#1-21,24,25
(and brief version)	3.5	17,18
Berresford-Rockett	7.3	#17-33, 39-44
Calculus with Finite Mathematics	7.4	#1-21,24,25
	7.5	17,18

Chapter 12

Exponential, Logarithmic and Trigonometric Functions

Excel has built-in functions for easily evaluating the exponential, logarithmic, and trigonometric functions. By using these built-in functions, you can easily create graphs and solve equations.

To enter a function $f(x)=ca^x$, you simply enter the formula in the same way as introduced in Chapter 1. The only exception is $f(x)=e^x$, which is entered as **`=exp(number)`**

For logarithmic functions of any base, you enter the formula **`=log(number,base)`**

The natural log function is entered as **`=ln(number)`** and the logarithm to base 10 is entered as **`=log10(number)`**

The following table summarizes all the syntax for the exponential and log functions, as well as the trig functions.

Function	Excel syntax
e^x	=exp(number)
ln(x)	=ln(number)
$\log_{10}(x)$	=log10(number)
$\log_a x$	=log(number,a)
sin(x)	=sin(number)
cos(x)	=cos(number)
tan(x)	=tan(number)
Arcsin(x)	=asin(number)
Arccos(x)	=acos(number)
Arctan(x)	=atan(number)

Graphs of Exponential, Log and Trig Functions

Example 1 Plot $f(x)=1.5^x$ and $g(x)=e^x$ on the interval [-2,3] on one plot. Which function grows more rapidly?

Solution

1 Make a table with three columns, following the directions for plotting more than one function in Chapter 2. Your table should be similar to the one below, shown in formula view.

	A	B	C
1	x	f(x)	g(x)
2	-2	=1.5^A2	=EXP(A2)
3	-1.5	=1.5^A3	=EXP(A3)
4	-1	=1.5^A4	=EXP(A4)
5	-0.5	=1.5^A5	=EXP(A5)
6	0	=1.5^A6	=EXP(A6)
7	0.5	=1.5^A7	=EXP(A7)
8	1	=1.5^A8	=EXP(A8)
9	1.5	=1.5^A9	=EXP(A9)
10	2	=1.5^A10	=EXP(A10)
11	2.5	=1.5^A11	=EXP(A11)
12	3	=1.5^A12	=EXP(A12)

2 Select the entire table and plot using ChartWizard. Your plot should be similar to the following.

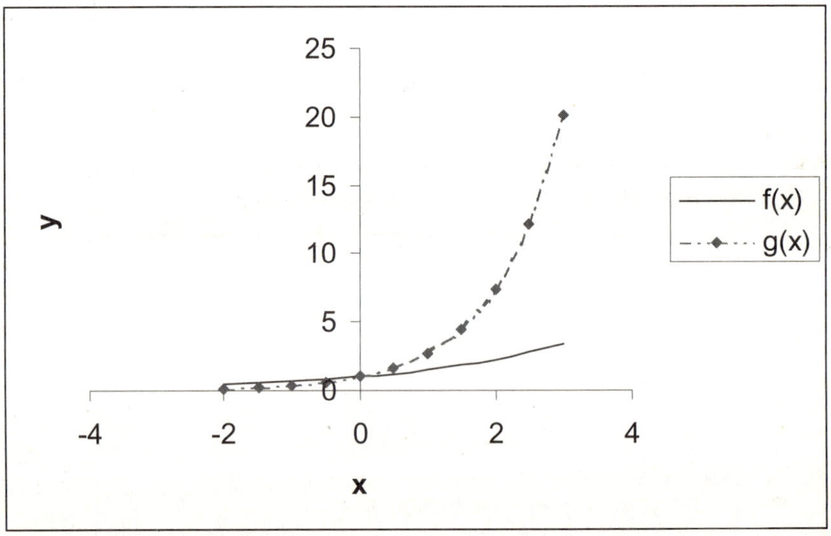

3 From the plot, it is easy to see that e^x increases more rapidly that 1.5^x.

Check it out

- Plot the graphs of f(x)=ln(x) and g(x)=$\log_{10}$(x) on the same plot using the interval [0.5,4]. Comment on the differences between the two graphs.

Example 2 Graph the functions $f(x) = \sin(x+\pi)$ and $g(x) = \sin(x)$ on the interval $[0, 2\pi]$.

Solution

1 First note that in Excel, the constant π is entered as `=pi()`

2 Make a table with three columns, following the directions for plotting more than one function in Chapter 2. Your ending value for x will be 6.5, if you increase in increments of 0.5. You may also stop at 6.0 and add 6.28 (approximately 2*pi) at the end of the table.

3 Your table should be similar to the one below, shown in formula view.

	A	B	C
1	x	f(x)	g(x)
2	0	=SIN(A2+PI())	=SIN(A2)
3	0.5	=SIN(A3+PI())	=SIN(A3)
4	1	=SIN(A4+PI())	=SIN(A4)
5	1.5	=SIN(A5+PI())	=SIN(A5)
6	2	=SIN(A6+PI())	=SIN(A6)
7	2.5	=SIN(A7+PI())	=SIN(A7)
8	3	=SIN(A8+PI())	=SIN(A8)
9	3.5	=SIN(A9+PI())	=SIN(A9)
10	4	=SIN(A10+PI())	=SIN(A10)
11	4.5	=SIN(A11+PI())	=SIN(A11)
12	5	=SIN(A12+PI())	=SIN(A12)
13	5.5	=SIN(A13+PI())	=SIN(A13)
14	6	=SIN(A14+PI())	=SIN(A14)
15	6.28	=SIN(A15+PI())	=SIN(A15)

4 Select the entire table and plot using ChartWizard. Your plot should be similar to the following.

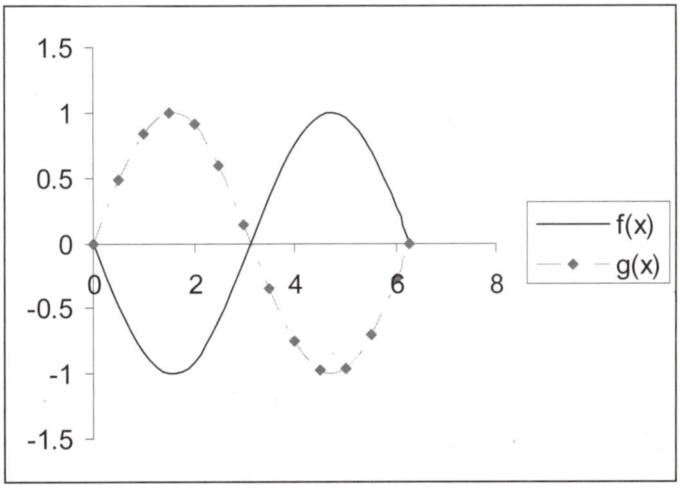

Check it out

- Graph f(x)=cos(x) and f(x)=2cos(x) on the interval [0,2*pi]. What observations can you make?

Suggested Exercises

Text	Section(s)	Exercises
Harshbarger-Reynolds *Mathematical Applications,* 6th ed.	5.1 5.2	#23-30, 39-51 #39-46, 63-67
Larson-Edwards *Calculus: An Applied Approach,* 6th ed. (and brief version)	4.1-4.4 8.1-8.6	All exercises using a graphing utility
Berresford-Rockett *Applied Calculus,* 3rd ed. (and brief version)	4.1-4.3 8.1-8.3	All exercises using a graphing utility
Berresford-Rockett *Calculus with Finite Mathematics*	8.1-8.3	All exercises using a graphing utility

Solving Equations with Goal Seek

Many of the exponential and logarithmic equations given in your textbook can be solved by using Goal Seek. Review Goal Seek in Chapter 4 before proceeding to the following example.

Example The number of daily sales after an advertising campaign declines according to the following model: $S = 1000(2^{-0.1x})$, where x is the number of days following the end of the campaign. When will the sales decline to 500 sales per day?

Solution

1 To solve this problem we must solve the equation $500 = 1000(2^{-0.1x})$.

2 This can be easily entered in Excel. The following table shows the setup.

	A	B
1	x	S
2	0	=1000*2^(-0.1*A2)

3 Invoke Goal Seek as explained in Chapter 4. After entering all the proper cell references, your dialog box should look like the following.

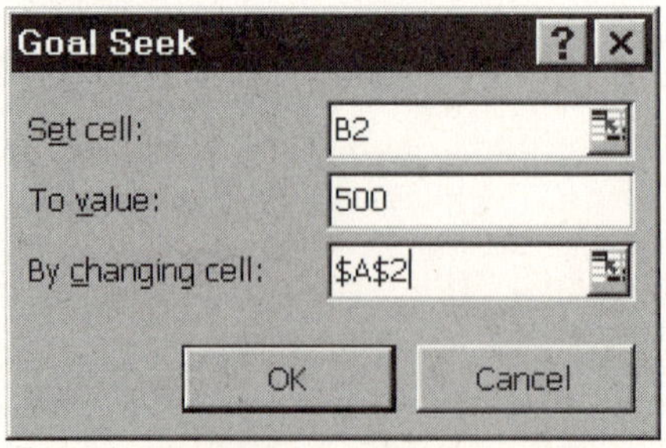

4 Click OK and the solution is computed. You should get a value of 10 for x. Hence it will take 10 days after the campaign end for sales to decline to 500. The solution on the spreadsheet is shown below.

	A	B
1	x	S
2	10	500.00029

Check it out

- How long will it take to reach 500 sales per day if the model is given by $S=1000e^{-0.1x}$?
- Make graphs of both models of S to confirm your answer.

Suggested Exercises

Text	Section(s)	Exercises
Harshbarger-Reynolds *Mathematical Applications,* 6th ed.	5.3	#1-45
Larson-Edwards *Calculus: An Applied Approach,* 6th ed. (and brief version)	4.5	#67-72
Berresford-Rockett *Applied Calculus,* 3rd ed. (and brief version)	4.4	#40,41
Berresford-Rockett *Calculus with Finite Mathematics*	8.4	#40,41

Chapter 13

Integration

Using the automatic features of table building and formula entry in Excel, we can easily approximate definite integrals by using various approximation methods. The first two sections will view the definite integral as the limit of a sum. The third and fourth sections will cover the Trapezoidal and Simpson's Rules to approximate a definite integral.

Approximating Area using Rectangles - Left Endpoint

The area under a curve f from x=a to x=b is denoted by the definite integral $\int_a^b f(x)dx$. This value can be approximated by the sum of the areas of a finite number of rectangles. We will assume that the width of each rectangle is the same. The width of each rectangle is then given by $\Delta x = \frac{(b-a)}{n}$, where n is the number of rectangles. The height of each rectangle is taken to be the height of the curve f at the left hand edge of the rectangle.

Example Approximate the area under the curve $f(x) = x^2$ from x=1 to 2 by five rectangles. Use rectangles with equal bases and with heights equal to the height of the curve at the left hand edge of the rectangles.

Solution

Steps to evaluate the sum of areas of rectangles

1 Since we are using five rectangles, the width of each rectangle will be $\Delta x = \frac{(2-1)}{5} = 0.2$.
2 The x-values of the left hand endpoints of each of the five rectangles will be x=1.0,1.2,1.4,1.6,1.8 .
3 Make a table containing the x-value of the left hand endpoint of each of the five rectangles in the first column. You can use the Autofill feature as explained in the Tables section in Chapter 1.
4 The second column of the table will contain the values of f(x) evaluated at the x-values in the first column. You should, of course, use formulas. Review the details in the Tables section in Chapter 1.
5 Your table will look like the following.

	A	B
1	x	f(x)
2	1	1
3	1.2	1.44
4	1.4	1.96
5	1.6	2.56

	A	B
6	1.8	3.24

6 Now add a third column with the heading "Delta x" in cell **C1**.

7 Enter the value 0.2 for Δx in cells **c2:c6**.

8 Add a fourth column heading "Area of each Rectangle" in cell **D1**.

9 The area of each rectangle is given by the formula $f(x_i)\Delta x$, where x_i is the value of each of the left hand endpoints. Therefore, in the cell **D2**, enter the formula **=B2*C2**

10 Copy this formula down to cells **D3:D6**. The total area is given by the sum of all the smaller rectangular areas.

- **a** In cell **C7**, type the heading "Total Area" .
- **b** In cell **D7**, enter the formula **=sum(D2:D6)** and press <ENTER>. Sum is a built-in Excel function which will sum all the entries in the cell references given in parentheses.

11 Your table should now look like the following.

	A	B	C	D	E	F
1	x	f(x)	Delta x	Area of each rectangle		
2	1	1	0.2	0.2		
3	1.2	1.44	0.2	0.288		
4	1.4	1.96	0.2	0.392		
5	1.6	2.56	0.2	0.512		
6	1.8	3.24	0.2	0.648		
7			Total area	2.04		

12 From the calculations, the area under the curve is approximately 2.04 square units.

Check it out

- Calculate the same area as in the above example, but now use ten rectangles, that is, $\Delta x = 0.1$. You should obtain 2.185 for the approximate area. Why is it closer to the actual value of 2 1/3 square units?
- Repeat with 100 rectangles. Your answer should be 2.318.

Approximating Area using Rectangles - Midpoint

In the previous section, we approximated the area under a curve f from x=a to x=b by using areas of rectangles whose heights were given by the height of the function at the left hand edge of the rectangle. We will assume that the width of each rectangle is the same. The width of each rectangle is then given by $\Delta x = \dfrac{(b-a)}{n}$, where n is the number of rectangles.A better approximation to the area can be obtained by using the height of the rectangle to be the value of the function at the midpoint of the subintervals of width Δx.

Example Approximate the area under the curve $f(x) = x^2$ from x=1 to 2 by five rectangles. Use rectangles with equal bases and with heights equal to the height of the curve evaluated at the midpoint of the bottom edge of each of the rectangles.

Solution

Steps to evaluate the sum of areas of rectangles

1 Since we are using five rectangles, the width of each rectangle will be $\Delta x = \frac{(2-1)}{5} = 0.2$.

2 The x-values of the midpoints of the bottom edges of each of the five rectangles will be x=1.1,1.3,1.5,1.7,1.9 .

3 Make a table containing the x-value of the midpoints of the bottom edges of each of the five rectangles in the first column. You can use the Autofill feature as explained in the Tables section in Chapter 1.

4 The second column of the table will contain the values of f(x) evaluated at the x-values in the first column. You should, of course, use formulas. Review the details in the Tables section in Chapter 1.

5 Your table will look like the following.

	A	B
1	x	f(x)
2	1.1	1
3	1.3	1.69
4	1.5	2.25
5	1.7	2.89
6	1.9	3.61

6 Now add a third column with the heading "Delta x" in cell **C1**.

7 Enter the value 0.2 for Δx in cells **c2:c6**.

8 Add a fourth column heading "Area of each Rectangle" in cell **D1**.

9 The area of each rectangle is given by the formula $f(x_i)\Delta x$, where x_i is the value of each of the midpoints. Therefore, in the cell **D2**, enter the formula **=B2*C2**

10 Copy this formula to cells **D3:D6**. The total area is given by the sum of all the smaller rectangular areas.

a In cell **C7**, type the heading "Total Area" .

b In cell **D7**, enter the formula **=sum(D2:D6)** and press <ENTER>. Sum is a built-in Excel function which will sum all the entries in the cell references given in parentheses.

11 Your table should now look like the following.

	A	B	C	D	E
1	x	f(x)	Delta x	Area of each rectangle	
2	1.1	1.21	0.2	0.242	
3	1.3	1.69	0.2	0.338	
4	1.5	2.25	0.2	0.45	
5	1.7	2.89	0.2	0.578	
6	1.9	3.61	0.2	0.722	
7			Total area	2.33	
8					

12 From the calculations, the area under the curve is approximately 2.330 square units.

Check it out

- Calculate the same area as in the above example, but now use ten rectangles, that is, $\Delta x = 0.1$. You should obtain 2.3325 for the approximate area. Why is it closer to the actual value of 2 1/3 square units?

Suggested Exercises

Text	Section(s)	Exercises
Harshbarger-Reynolds *Mathematical Applications, 6th ed.*	13.1	#1-8, #31,32
Larson-Edwards *Calculus: An Applied Approach, 6th ed.* (and brief version)	5.6	#13-16, 27, 28, 33
Berresford-Rockett *Applied Calculus, 3rd ed.* (and brief version)	5.3 5.4	#9-18 #31,32
Berresford-Rockett *Calculus with Finite Mathematics*	9.3 9.4	#3-12 #31,32

Numerical Integration - Trapezoidal Rule

Many integrals cannot be evaluated exactly. Therefore, numerical methods must be used to approximate the definite integral. While the rectangular rules discussed above may be used to approximate the definite integral, they are not usually accurate enough.

One method which gives good results and is easy to implement is the Trapezoidal Rule. The formula for the trapezoidal rule is

$$\int_a^b f(x)dx \approx \frac{b-a}{2n}(f(x_0) + 2f(x_1) + \ldots + 2f(x_{n-1}) + f(x_n))$$

where $x_0, x_1, \ldots, x_n$ are equally spaced points on the interval $[a, b]$ with $x_0 = a$ and $x_n = b$.

Example Approximate $\int_0^1 e^x dx$ using the Trapezoidal rule with n=4 subintervals.

Solution

Steps to implementing the trapezoidal rule

1 Since n=4, $\Delta x = \frac{(1-0)}{4} = 0.25$.

2 The x-values for which the trapezoidal rule will be evaluated will be x=0,0.25,0.5,0.75,1.

3 Make a heading for the x-values in Step 2 by typing "x" in cell **A1**.
4 Fill in the x-values in **a2:a6**. You can use the Autofill feature as explained in the Tables section in Chapter 1.
5 The second column of the table will contain the values of f(x) evaluated at the x-values in the first column. You should, of course, use formulas. To enter the exponential function in Excel, type the formula **=exp(a2)** in cell **b2**. Copy the formula to cells **b3:b6**.
6 Your table will look like the following.

	A	B
1	x	f(x)
2	0	1
3	0.25	1.284
4	0.5	1.648
5	0.75	2.117
6	1.0	2.718

7 Now add a third column with the heading "Weight" in cell **C1**.
8 The first and last f(x)-values are multiplied by 1 in the formula. Hence, they will have 1 typed into the cells **c2** and **c6**.
9 All the other values of f(x) are multiplied by 2. Therefore, type 2 into cells **c3:c5**. your table should look like the following (shown in formula view).

	A	B	C
1	x	f(x)	Weight
2	0	=EXP(A2)	1
3	0.25	=EXP(A3)	2
4	0.5	=EXP(A4)	2
5	0.75	=EXP(A5)	2
6	1	=EXP(A6)	1
7			

10 Type the heading "Weight*f(x)" in cell **D1**. The column D will contain the product of the weight and f(x). Therefore, in the cell **D2**, enter the formula **=B2*C2**
11 Copy this formula to cells **D3:D6.**
12 The total area using the trapezoidal rule is given by the sum of the entries in **d2:d6** multiplied by

$$\frac{b-a}{2n} = \frac{1-0}{(2)(4)} = 0.5\Delta x \ .$$

a In cell **A9**, type the heading "Delta x".
b In cell **A10**, type 0.25 as the value for Δx
c In cell **A11**, type the heading "Total Area".
d In cell **A12**, enter the formula **=0.5*A10*sum(D2:D6)** and press <ENTER>. Sum is a built-in Excel function which will sum all the entries in the cell references given in parentheses.

13 Your table should now look like the following.

	A	B	C	D	E
1	x	f(x)	Weight	Weight*f(x)	
2	0	1	1	1	
3	0.25	1.284025	2	2.56805083	
4	0.5	1.648721	2	3.29744254	
5	0.75	2.117	2	4.23400003	
6	1	2.718282	1	2.71828183	
7					
8					
9	Delta x				
10	0.25				
11	Total Area				
12	1.7272219				
13					

14 From the calculations, the area under the curve is approximately 1.727square units.

Numerical Integration - Simpson's Rule

An integration rule which is more accurate than the trapezoidal rule is Simpson's Rule. The formula for the rule is, for n even,

$$\int_a^b f(x)dx \approx \frac{b-a}{3n}[f(x_0) + 4f(x_1) + 2f(x_2) + 4f(x_3) + \ldots + 4f(x_{n-1}) + f(x_n)]$$

where $x_0, x_1, \ldots, x_n$ are equally spaced points on the interval $[a, b]$ with $x_0 = a$ and $x_n = b$. Note that the coefficients of Simpson's Rule have the pattern

1 4 2 4 2 4 ... 4 2 4 1

Example Approximate $\int_0^1 e^x dx$ using Simpson's rule with n=4 subintervals.

Solution

Steps to implementing Simpson's rule

1 Since n=4, $\Delta x = \frac{(1-0)}{4} = 0.25$.

2 The x-values for which the Simpson's rule will be evaluated will be x=0,0.25,0.5,0.75,1.
3 Make a heading for the x-values in Step 2 by typing "x" in cell `A1`.
4 Fill in the x-values in `a2:a6`. You can use the Autofill feature as explained in the Tables section in Chapter 1.

5 The second column of the table will contain the values of f(x) evaluated at the x-values in the first column. You should, of course, use formulas. To enter the exponential function in Excel, type the formula `=exp(a2)` in cell **b2**. Copy the formula to cells **b3:b6**.

6 Your table will look like the following.

	A	B
1	x	f(x)
2	0	1
3	0.25	1.284
4	0.5	1.648
5	0.75	2.117
6	1.0	2.718

7 Now add a third column with the heading "Weight" in cell **C1**.

8 The first and last f(x)-values are multiplied by 1 in the formula. Hence, they will have 1 typed into the cells **c2** and **c6**.

9 All the other values of f(x) are multiplied by the values 4 and 2, alternately. To automatically have Excel enter these into the cells, you will need to use an **If** statement as follows:

- a Enter the number 4 in cell **c3**, since that is what f(0.25) is multiplied by.
- b In cell **c4**, enter the formula `=if(c3=4,2,4)`
- c The above formula simply checks if the entry above is equal to 4. If it is, the current entry will be assigned a value of 2. Otherwise, it will be assigned a value of 4. This will automatically generate the alternating values of 4 and 2. When you have a large number of intervals, this method is much faster.
- d Copy the formula in **c4** to **c5**. Your table will look like the following (shown in formula view).

	A	B	C
1	x	f(x)	Weight
2	0	=EXP(A2)	1
3	0.25	=EXP(A3)	4
4	0.5	=EXP(A4)	=IF(C3=4,2,4)
5	0.75	=EXP(A5)	=IF(C4=4,2,4)
6	1	=EXP(A6)	1
7			

10 Type the heading "Weight*f(x)" in cell **D1**. The column D will contain the product of the weight and f(x). Therefore, in the cell **D2**, enter the formula **=B2*C2**

11 Copy this formula to cells **D3:D6**.

12 The total area using Simpson's rule is given by the sum of the entries in **d2:d6** multiplied by

$$\frac{b-a}{3n} = \frac{1-0}{(3)(4)} = \left(\frac{1}{3}\right)\Delta x \ .$$

- a In cell **A9**, type the heading "Delta x".
- b In cell **A10**, type in the value of $\Delta x = 0.25$
- c In cell **A11**, type the heading "Total Area".

d In cell **`A12`**, enter the formula **`=(1/3)*A10*sum(D2:D6)`** and press <ENTER>. Sum is a built-in Excel function which will sum all the entries in the cell references given in parentheses.

13 Your table should now look like the following.

	A	B	C	D
1	**x**	**f(x)**	**Weight**	**Weight*f(x)**
2	0	1	1	1
3	0.25	1.284	4	5.13610167
4	0.5	1.649	2	3.29744254
5	0.75	2.117	4	8.46800007
6	1	2.718	1	2.71828183
7				
8				
9	**Delta x**			
10	0.25			
11	**Total Area**			
12	1.7183188			

14 From the calculations, the area under the curve is approximately 1.718 square units. Note that this value is more accurate than the one given by the trapezoidal rule.

Suggested Exercises

Text	Section(s)	Exercises
Harshbarger-Reynolds *Mathematical Applications, 6th ed.*	N/A	
Larson-Edwards *Calculus: An Applied Approach, 6th ed.* (and brief version)	6.5	#21-46
Berresford-Rockett *Applied Calculus, 3rd ed.* (and brief version)	6.4	#15-18,27-36
Berresford-Rockett *Calculus with Finite Mathematics*	10.4	#9-33

Chapter 14

Graphs of Functions of Two Variables

We can create surface plots of functions of two variables in Excel in a manner similar to creating graphs of functions of one variable. Keep in mind that now we should have a set of x and y values so that we can generate a table of values for z=f(x,y). The next example shows how to do this.

Example Plot the graph of $f(x,y) = 10-x^2-y^2$ for (x,y) in [-2,2]X[-2,2].

Solution

Steps to generating a table

1 Generate x values from -2 to 2 in increments of 0.5, beginning in cell **B1** and continuing *across*.
2 Generate y-values from -2 to 2 in increments of 0.5, beginning in cell **A2** and continuing *down.* A view of the table is given below.

	A	B	C	D	E	F	G	H	I	J
1		-2	-1.5	-1	-0.5	0	0.5	1	1.5	2
2	-2									
3	-1.5									
4	-1									
5	-0.5									
6	0									
7	0.5									
8	1									
9	1.5									
10	2									
11										

3 The next step is to generate values of f(x,y), where the x and y values correspond to the x and y values generated in steps 1 and 2.

- **a** In cell **B2**, enter the formula for f(x,y) with x corresponding to the x-value in cell **B1** and y corresponding to the y-value in cell **A2**: **=10-B1^2-A2^2**
- **b** In the formula above, the reference **b$1** states that the row number should be fixed, but the column number should vary when copying the formula across. Thus, we vary through all the x-values in the topmost row.
- **c** Similarly, the reference **$A2** states that the column number should be fixed, but the row number should vary. Thus, when we copy the formula across row 1, x varies but y is held fixed.
- **d** Copy the formula in **b2** across Row 1.
- **e** A partial view of the first row of the f(x,y) table, in formula view, is shown below.

	A	B	C	D	E	F
1		-2	-1.5	-1	-0.5	0
2	-2	=10-B1^2-A2^2	=10-C1^2-A2^2	=10-D1^2-A2^2	=10-E1^2-A2^2	=10-F1^2-A2^2
3	-1.5					

- **f** Carefully look at the formulas in the table above to make sure you understand where the values of x and y come from.

g Highlight the formulas in row from **b2:j2** and copy down to fill up the rest of table. A partial view of the table, in formula view, is given below.

	A	B	C	D	E
1		-2	-1.5	-1	-0.5
2	-2	=10-B1^2-A2^2	=10-C1^2-A2^2	=10-D1^2-A2^2	=10-E1^2-A2^2
3	-1.5	=10-B1^2-A3^2	=10-C1^2-A3^2	=10-D1^2-A3^2	=10-E1^2-A3^2
4	-1	=10-B1^2-A4^2	=10-C1^2-A4^2	=10-D1^2-A4^2	=10-E1^2-A4^2
5	-0.5	=10-B1^2-A5^2	=10-C1^2-A5^2	=10-D1^2-A5^2	=10-E1^2-A5^2
6	0	=10-B1^2-A6^2	=10-C1^2-A6^2	=10-D1^2-A6^2	=10-E1^2-A6^2
7	0.5	=10-B1^2-A7^2	=10-C1^2-A7^2	=10-D1^2-A7^2	=10-E1^2-A7^2
8	1	=10-B1^2-A8^2	=10-C1^2-A8^2	=10-D1^2-A8^2	=10-E1^2-A8^2
9	1.5	=10-B1^2-A9^2	=10-C1^2-A9^2	=10-D1^2-A9^2	=10-E1^2-A9^2
10	2	=10-B1^2-A10^2	=10-C1^2-A10^2	=10-D1^2-A10^2	=10-E1^2-A10^2

h Your table of values as shown below.

	A	B	C	D	E	F	G	H	I	J
1		**-2**	**-1.5**	**-1**	**-0.5**	**0**	**0.5**	**1**	**1.5**	**2**
2	**-2**	2	3.75	5	5.75	6	5.75	5	3.75	2
3	**-1.5**	3.75	5.5	6.75	7.5	7.75	7.5	6.75	5.5	3.75
4	**-1**	5	6.75	8	8.75	9	8.75	8	6.75	5
5	**-0.5**	5.75	7.5	8.75	9.5	9.75	9.5	8.75	7.5	5.75
6	**0**	6	7.75	9	9.75	10	9.75	9	7.75	6
7	**0.5**	5.75	7.5	8.75	9.5	9.75	9.5	8.75	7.5	5.75
8	**1**	5	6.75	8	8.75	9	8.75	8	6.75	5
9	**1.5**	3.75	5.5	6.75	7.5	7.75	7.5	6.75	5.5	3.75
10	**2**	2	3.75	5	5.75	6	5.75	5	3.75	2

Steps to graphing a function of two variables

1. Select the entire table from **a1:j10**.
2. Click on the Chartwizard icon.

3 Choose Surface plot with the first chart subtype as shown below.

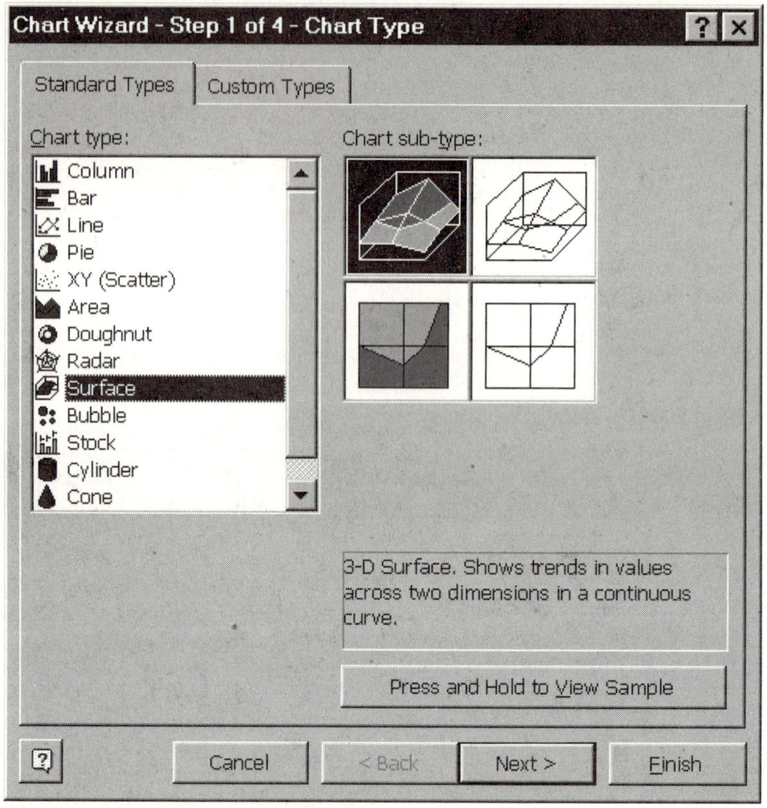

4 Click Next.
5 Also, click Next in Step 2.
6 Annotate your graph in Step 3 and click Finish.
7 Your graph should resemble the one below. It is shaded according to the z-values.

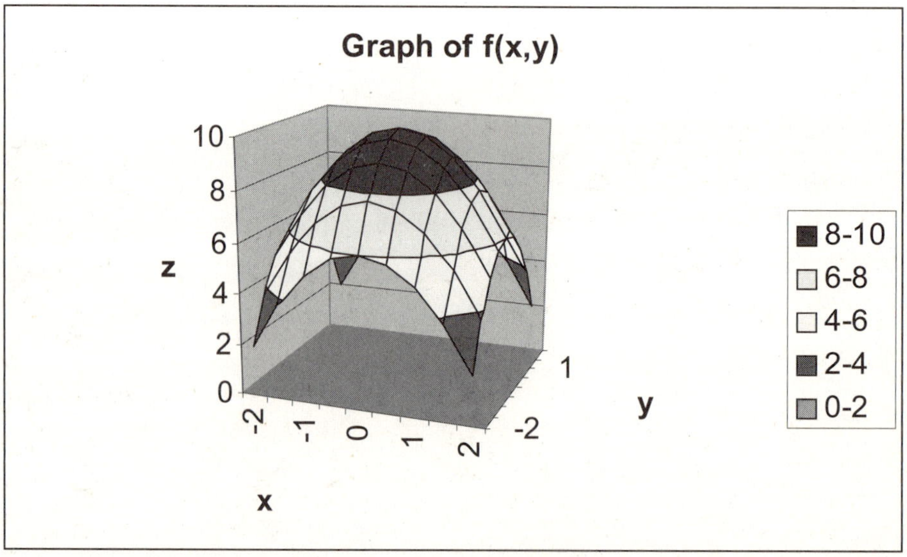

8 You can click into the graph and move it around to see from a different perspective.

Check it out

- Plot the graph of $f(x,y)=x^2+2y^2$ on the interval [-3,3]X[-2,2].

Suggested Exercises

Text	Section(s)	Exercises
Harshbarger-Reynolds *Mathematical Applications, 6th ed.*	14.1	Graph functions in #1-8
Larson-Edwards *Calculus: An Applied Approach, 6th ed.* (and brief version)	7.3	Graph functions in #1-7
Berresford-Rockett *Applied Calculus, 3rd ed.* (and brief version)	7.1	Graph the functions in the gallery of surfaces
Berresford-Rockett *Calculus with Finite Mathematics*	11.1	Graph the functions in the gallery of surfaces

Chapter 15

Constrained Optimization and Lagrange Multipliers

This chapter will illustrate the use of an Excel tool called Solver to solve constrained optimization-problems. To check that your installation of Excel has Solver, select Tools in the menu bar. If you see Solver as one of the options, you are ready to go. Otherwise, you will have to add it in. See the preface on how to add in Solver.

Example A company's output is given by the Cobb-Douglas production function $P = 600l^{\frac{2}{3}}k^{\frac{1}{3}}$, where l and k are the number of units of labor and capital. Each unit of labor costs the company $40 and each unit of capital costs $100. If the company has a total of $3000 for labor and capital, how much of each should it use to maximize production?

Solution

Set up of problem

1 Identify variables

l: units of labor

k: units of capital

2 Identify objective: Maximize the production function $P = 600l^{\frac{2}{3}}k^{\frac{1}{3}}$

3 The objective function is subject to the following constraint on labor and capital:

$$40l + 100k = 3000$$

The next step is to input all this information into Excel so that Solver can be invoked. Since all the calculations in the spreadsheet are done with cell references, you must set up cell entries for the variables, objective function and constraint. In Excel, the cell containing the formula for the objective function is referred to as the *target cell*. The cells containing the variables are called the *changing cells*. The constraints are referred to as, well, *constraints*.

Steps to set up the problem in Excel

1 In a blank spreadsheet, first type a heading called "Variables" in cell `a1`, followed by the variable descriptions in `a3:a4` and values in cells `b3:b4`. The variables are initially assigned values of zero. Refer to the table below as a guide.

2 The objective function formula is given in terms of the cell references for the variables l and k. Enter the information for the objective function as follows:

- **a** Type a heading called "Objective" in cell **`a7`**
- **b** Type a description of the objective in cell **`a9`**
- **c** Enter the objective function formula in **`b9`** .The formula is **`=600*b3^(2/3)*b4^(1/3)`**

3 Type in the formulas for the constraint.

- **a** Type a heading called "Constraint" in **`a11`** and a descriptive label in **`a13`**.
- **b** The formula for the constraint is also given in terms of the cell references for l and k and is contained in **`b13`** .
- **c** The amount of labor and capital available is typed in **`c13`**. The complete setup of formulas and other entries is shown below.

	A	B	C
1	**Variables**		
2			
3	units labor (l)	0	
4	units capital (k)	0	
5			
6			
7	**Objective**		
8			
9	Maximize production	=600*B3^(2/3)*B4^(1/3)	
10			
11	**Constraint**		
12		Amount used	Available
13	Cost	=40*B3+100*B4	3000

Check it out

- To get familiar with the setup of the problem in Excel, change the variables in **`b3:b4`** to some nonzero values. What happens to the value of the objective function? What happens to the value for the constraint?

Steps to solve the problem using Solver

1 Once you check that your spreadsheet contains all the correct formulas in the appropriate cells, you are ready to invoke Solver by choosing **Tools > Solver**.

2 You will see a dialog box whose first entry is the information for the target cell (i.e. the objective function). Click cursor into the this entry box and click into cell **`B9`** (formula for objective function).

3 Check the button to maximize. Your dialog box should now look like the one below.

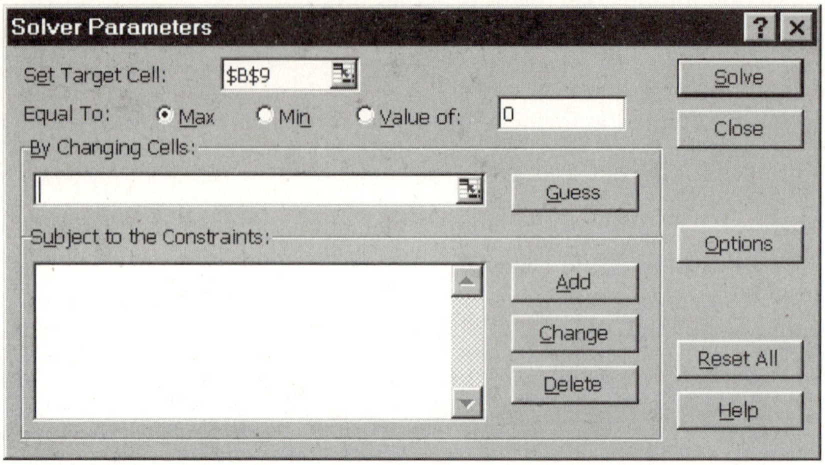

4 Next Click cursor to the By Changing Cells entry box.

5 Enter the cell references for the variables by selecting the cells **`b3:b4`** . This portion of your dialog box should resemble the following.

6 Now you will add the constraint.

- **a** Click cursor into Subject to the Constraints entry box.
- **b** Press the Add button to add the first constraint. You will get a new dialog box for the constraint.
- **c** Click cursor to the left entry box and click into cell **`b13`** containing the formula for the first constraint.
- **d** The middle entry box should be set to =.
- **e** Click cursor to the right entry box and click into the cell **`c13`** containing the available quantity. Your constraint dialog box should resemble the following.

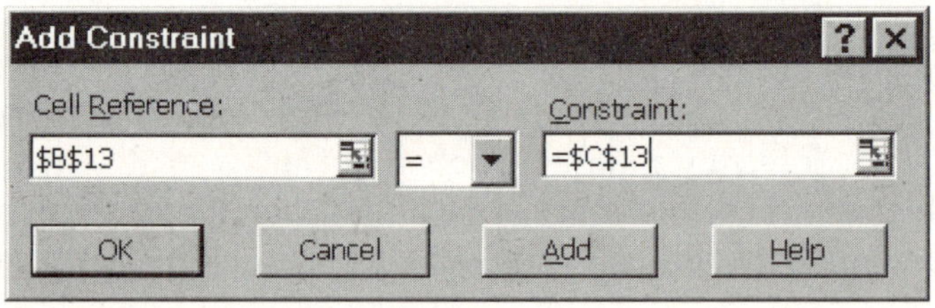

- **f** Click OK.

7 Your completed Solver box should resemble the following.

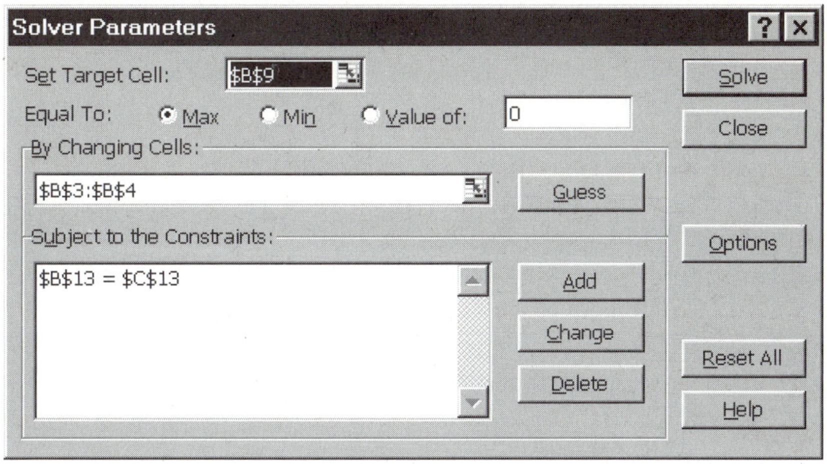

8 Click into the Options box, and make sure that the Assume Linear Model checkbox is NOT checked, as in the following figure.

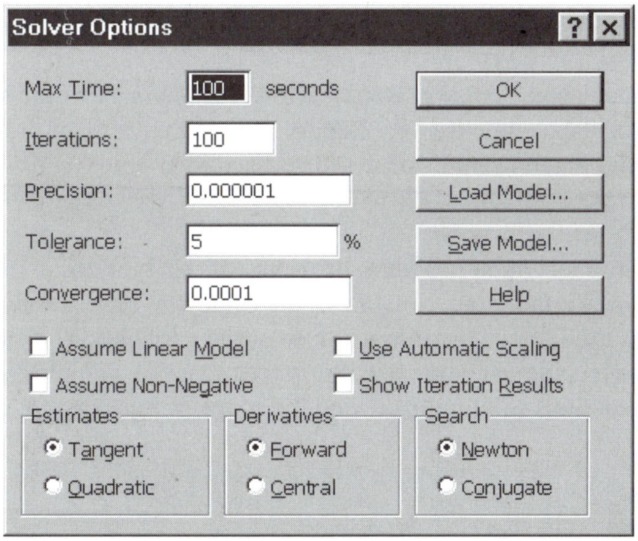

9 Click OK in the Options dialog box.
10 Click Solve in the Solver dialog box. You will get a dialog box stating that Solver found a solution.
11 Check the Keep Solver Solution button and also select the Answer and Sensitivity reports. Your completed box resembles the following

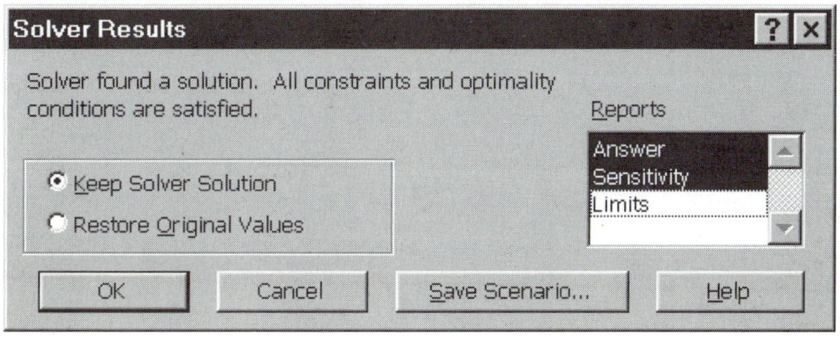

12 Click OK. Go back and examine the cells with the variables, constraints, and objective. They should now contain the optimal values and resemble the following table.

	A	B	C
1	**Variables**		
2			
3	units labor (l)	50.00000002	
4	units capital (k)	10	
5			
6			
7	**Objective**		
8			
9	Maximize production	17544.10644	
10			
11	**Constraint**		
12		Amount used	Available
13	Cost	3000.000001	3000

13 From the results above, we see that 50 units of labor and 10 units of capital are needed to maximize production.

14 Click on the worksheet labeled Answer Report 1 to see a summary of the solution.

15 Click on the worksheet labeled Sensitivity Report 1 to see the value of the Lagrange multiplier. The significance of the Lagrange multiplier is discussed in your textbook.

Note: This is a nonlinear optimization problem. Problems of this type are in general more difficult to solve than linear problems. Furthermore, Solver can only find one set of solutions at a time. It usually finds the solution set that is closest to the initial values of the variables. You should always examine the answer given by any computer solver to see if it makes sense. A complete discussion of nonlinear optimization and their numerical solution is beyond the scope of this manual.

Suggested Exercises

Text	Section(s)	Exercises
Harshbarger-Reynolds *Mathematical Applications, 6th ed.*	14.5	#17-20
Larson-Edwards *Calculus: An Applied Approach, 6th ed.* (and brief version)	7.6	#35-47, 49, 50
Berresford-Rockett *Applied Calculus, 3rd ed.* (and brief version)	7.5	#25-40
Berresford-Rockett *Calculus with Finite Mathematics*	11.5	#25-40

Chapter 16

Taylor Series

This chapter will show you how to use Excel to study the Taylor polynomial of a given function. Note that since Excel does not have any symbolic capabilities, you must find the derivatives by hand before you input the information into the spreadsheet.

Example Find the second and third degree Taylor polynomials of $f(x) = e^{-x}$ about x=0. Graph *f* and its second and third degree Taylor polynomials on the interval [-2,2].

Solution The second degree Taylor polynomial of $f(x) = e^{-x}$, denoted by $p_2(x)$, is given by

$$p_2(x) = f(0) + (x-0)f^{(1)}(0) + \frac{(x-0)^2}{2}f^{(2)}(0).$$

Steps to calculating and graphing $p_2(x)$

1 Type the heading "c" in cell **A1** to indicate the point of expansion.
2 Enter the value of c, the point you are expanding about, in cell **B1**. For this example set the value in **B1** to 0.
3 Type the headings f(0),f'(0), and f''(0) in cells , **A2**,**B2**, and **C2**, respectively.
4 Calculate f(0): in cell **A3**, enter the formula `=exp(-b1)`
5 Calculate f'(0): in cell **B3**, enter the formula `=-exp(-b1)`
6 Calculate f''(0): in cell **C3**, enter the formula `=exp(-b1)`
7 To start evaluating the polynomial at various x-values, first type the heading "x" in cell **A5**.
8 Make table of values for x in [-2,2] in steps of 0.25, starting in cell **A6**. You may refer to the Tables section in Chapter 1 to see how to do this.
9 Type the heading "f(x)" in cell **B5**.
10 In cell B6, type the formula `=exp(-a6)`
11 Copy this formula down to correspond to all x in the first column.
12 Type the heading "p_2(x)" in cell **C5**
13 Now enter the formula for p2(x): in cell **C6**, type the formula `=$A$3+(A6-$B$1)*($B$3)+(A6-$B$1)^2*($C$3)/2`

   a Explanation: **$A$3** denotes the value of f(0). **$B$3** and **$C$3** denote f'(0) and f''(0), respectively.
   b Explanation: **A6** stands for the first x-value
   c Explanation: **$B$1** contains the value for c, the point you are expanding about
   d Explanation: Using **$A$3** instead of **A3** in the above formula means that you can copy down the column without the changing the reference **A3**. Similarly for **$B$1**, **$B$3** and **$C$3**.

14 Copy the formula in cell **C6** down the column to correspond to all x in the first column.

15 Select the table and graph using X-Y Scatter with smooth curve -no markers option. See Chapter 2 on for more details on graphing. Your graph should resemble the following.

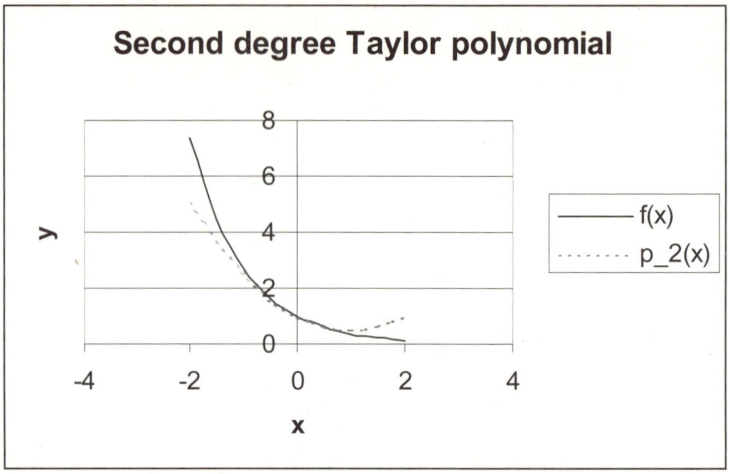

16 From the graph above, the polynomial approximation is best at points close to 0. Note how the approximation is not as good near x=2 and x=-2.

Check it out

- Change the value of c in cell B1 to -1. This will recalculate the Taylor polynomial centered at x=-1. What do you observe on the graph?
- Experiment with other values in [-2,2]. What do you observe?

Steps to calculating $p_3(x)$

The third degree Taylor polynomial of $f(x) = e^{-x}$, denoted by $p_3(x)$, is given by

$$p_3(x) = f(0) + (x-0)f^{(1)}(0) + \frac{(x-0)^2}{2}f^{(2)}(0) + \frac{(x-0)^3}{6}f^{(3)}(0).$$

$$= p_2(x) + \frac{(x-0)^3}{6}f^{(3)}(0)$$

1 Using the same worksheet as for calculating $p_2(x)$, add another heading f'''(0) in cell **D2**.
2 Enter the formula **`=-exp(-b1)`** in cell **D3** .
3 Enter the heading "p_3(x)" in cell **D5**.
4 Now enter the formula for $p_3(x)$: in cell **D6**, type the formula **`=c6+(a6-$b$1)^3*$d$3/6`**
5 In the formula above, we have simply added the value of p_2(x), in cell **c6**, to the last term for p_3(x).
6 Copy the formula in cell **D6** down the column to correspond to all x in the first column.

7 Select the table and graph using X-Y Scatter with smooth curve -no markers option. See Chapter 2 on for more details on graphing. Your graph should resemble the following.

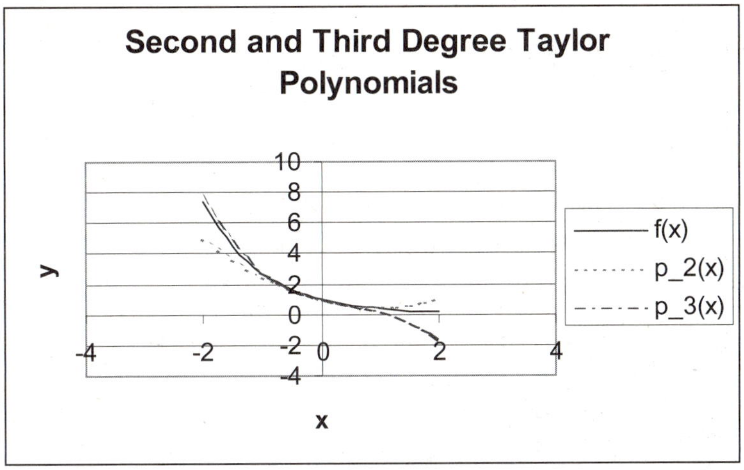

8 Note that the third degree Taylor polynomial provides a better approximation than the second degree Taylor polynomial. On this interval, f(x) and p_3(x) are almost the same.

Suggested Exercises

Text	Section(s)	Exercises
Larson-Edwards *Calculus: An Applied Approach,* 6th ed.	10.5	#7-8,13-16
Berresford-Rockett *Applied Calculus,* 3rd ed.	10.2 10.3	#9-12,17,18 #35,36

Chapter 17

Newton's Method

In calculus and other applications, it is often necessary to solve equations of the form $f(x) = 0$. Values of x which satisfy this equation are called zeros, or roots, of f. In many instances, these values cannot be calculated exactly and so approximating procedures must be used for finding the root. One such method is known as Newton's Method.

Finding a Zero of a Function

Example Find the zeros of the function $f(x) = x^2 - 6x + 7$.

Solution Before implementing the formula for Newton's Method, it is advisable to graph the function to see roughly where the root(s) of the function are. Newton's Method usually (but not always) finds the root closest to the initial guess input by the user. Follow the steps below to find one of the zeros of $f(x)$.

Steps to find one zero of a quadratic function

1 The vertex of the parabola is at x=-b/2a=3. Therefore, we pick an interval of x-values around x=3. For this example,we choose the interval [0,6]

2 Make a table of x and y values using the directions for tables in Chapter 1.

	A	B
1	x	f(x)
2	0	7
3	1	2
4	2	-1
5	3	-2
6	4	-1
7	5	2
8	6	7

3 Select the range of cells **A1:B8** and graph using the directions in Chapter 2. Your graph should look like the following:

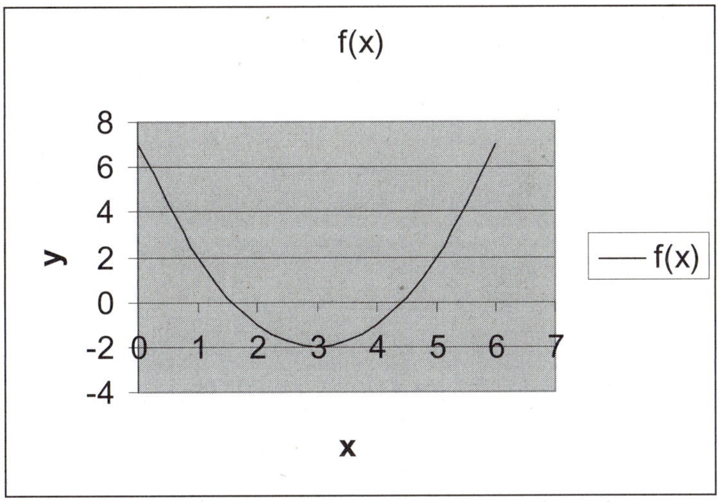

4 We see that there is one x-intercept near 2 and another near 4.

5 We next implement Newton's Method to find the root near x=2 by creating a table with headings as follows:

	A	B	C	D	E	F
1	n	x_n	f(x_n)	f'(x_n)	f(x_n)/f'(x_n)	x_n-f(x_n)/f'(x_n)
2						

6 The initial guess for the root near x=2 can be taken as $x_0=2$, with n=0. Using formulas, we evaluate $f(x_0)$, $f'(x_0)$, $f(x_0)/f'(x_0)$, and the next iterate, given by $x_0-f(x_0)/f'(x_0)$:

	A	B	C	D	E	F
1	n	x_n	f(x_n)	f'(x_n)	f(x_n)/f'(x_n)	x_n-f(x_n)/f'(x_n)
2	0	2	=B2^2-6*B2+7	=2*B2-6	=C2/D2	=B2-E2

7 Note how the formulas are entered. Recall that $f(x) = x^2 - 6x + 7$, f'(x)=2x-6.

8 The value in cell F2 is the next iterate x1. To use x1 as the next iterate, type 1 in cell **A3** and the formula **=f2** in cell **b3**

9 The results are as follows:

	A	B	C	D	E	F
1	n	x_n	f(x_n)	f'(x_n)	f(x_n)/f'(x_n)	x_n-f(x_n)/f'(x_n)
2	0	2	-1	-2	0.5	1.5
3	1	1.5				

10 Next, copy the formulas in **c2:f2** to **c3:f3** to get the following:

	A	B	C	D	E	F
1	n	x_n	f(x_n)	f'(x_n)	f(x_n)/f'(x_n)	x_n-f(x_n)/f'(x_n)
2	0	2	-1	-2	0.5	1.5
3	1	1.5	0.25	-3	-0.08333333	1.583333333

11 Type 2 in cell **A4** and copy the formulas in cells **b3:f3** to **b4:f4**.

12 Repeat above step two more times, adjusting the value of n accordingly. Your formulas should be as follows:

	A	B	C	D	E	F
1	n	x_n	f(x_n)	f'(x_n)	f(x_n)/f'(x_n)	x_n-f(x_n)/f'(x_n)
2	0	2	=B2^2-6*B2+7	=2*B2-6	=C2/D2	=B2-E2
3	1	=F2	=B3^2-6*B3+7	=2*B3-6	=C3/D3	=B3-E3
4	2	=F3	=B4^2-6*B4+7	=2*B4-6	=C4/D4	=B4-E4
5	3	=F4	=B5^2-6*B5+7	=2*B5-6	=C5/D5	=B5-E5
6	4	=F5	=B6^2-6*B6+7	=2*B6-6	=C6/D6	=B6-E6
7						

13 The actual values are as follows:

	A	B	C	D	E	F
1	n	x_n	f(x_n)	f'(x_n)	f(x_n)/f'(x_n)	x_n-f(x_n)/f'(x_n)
2	0	2	-1	-2	0.5	1.5
3	1	1.5	0.25	-3	-0.08333333	1.583333333
4	2	1.583333333	0.0069	-2.83	-0.00245098	1.585784314
5	3	1.585784314	6E-06	-2.83	-2.1239E-06	1.585786438
6	4	1.585786438	5E-12	-2.83	-1.5946E-12	1.585786438

14 Note that the values of x_3 and x_4 agree to within 5 decimal places. We can therefore take the value of x_4 to be an approximate root of $f(x) = x^2 - 6x + 7$ near x=2.

Check it out

- Use Newton's Method to find the root near x=4 in the above example. Your answer should be approximately 4.41421

Suggested Exercises

Text	Section(s)	Exercises
Larson-Edwards *Calculus: An Applied Approach,* 6th ed.	10.6	#17-28,42-44
Berresford-Rockett *Applied Calculus,* 3rd ed.	10.4	#11-20

Finding Intersection of Two Functions and Stopping Criteria

Newton's Method can also be applied to find the intersection of the graphs of two functions. This is useful in calculating quantities such as the break-even point for cost and revenue functions or the equilibrium point for supply and demand functions.

Example Approximate the solution to $e^x = 2 - 2x$, continuing until two successive iterations agree to within a tolerance of 10^{-8}. (Variation of Example 3, Sec. 10.4, Berresford-Rockett.)

Solution We proceed by first graphing the two functions on a suitable interval. Note that at x=0, $e^0 <$ 2-2(0). But at x=1, $e^1 >$ 2-2(1). Hence, there must be a point between 0 and 1 where the two functions are equal.

Steps to finding the intersection

1 Graph the functions e^x and 2-2x on the same plot on the interval [0,1] using subdivisions of 0.1 . This is discussed in Chapter 2 under the section *Graphing More than One Function*. Use the Excel function **exp** for the exponential function. Your plot should resemble the following.

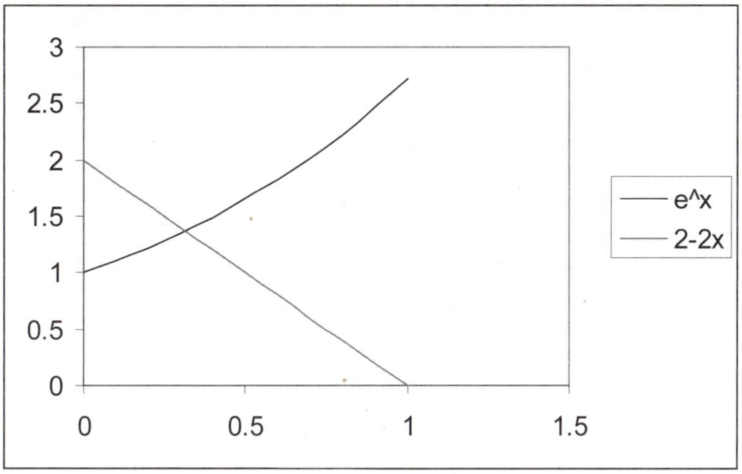

2 We see from the graph that there is an intersection between x=0 and x=1.

3 To use Newton's Method, we rewrite the equation $e^x = 2-2x$ as $e^x - 2 + 2x = 0$.

4 Apply Newton's Method to $f(x) = e^x - 2 + 2x$. Set up the table as in the previous example, with the new formulas for f(x) and $f'(x) = e^x + 2$.

	A	B	C	D	E	F
1	n	x_n	f(x_n)	f'(x_n)	f(x_n)/f'(x_n)	x_n-f(x_n)/f'(x_n)
2	0	0	=EXP(B2)-2+2*B2	=EXP(B2)+2	=C2/D2	=B2-E2

5 In cell **b3**, Type the formula **=f2** . Copy the formulas in **c2:e2** to cells **c3:e3**.

Stopping Criteria

1 Instead of simply generating iterates, we can automatically insert a stopping criteria in cell **F3**. It will check if the absolute value of the difference of the current iterate and the previous iterate is less than the tolerance specified. It will print "stop" if it is true. Otherwise, it will print the value of the current iterate.

2 In cell F3, type the formula **=IF(ABS((B3-E3)-B3) < 10^(-9), "Stop",B3-E3)**

3 Copy the formulas in `b3:f3` to `b4:f4`. Repeat with successive rows until you meet the stopping criterion. Your table will look like the following:

	A	B	C	D	E	F
1	n	x_n	f(x_n)	f'(x_n)	f(x_n)/f'(x_n)	x_n-f(x_n)/f'(x_n)
2	0	0	-1	3	-0.333333333	0.333333333
3	1	0.333333333	0.06228	3.39561243	0.018341048	0.314992285
4	2	0.314992285	0.00023	3.37024874	6.92261E-05	0.314923059
5	3	0.314923059	3.3E-09	3.37015389	9.74204E-10	Stop

4 From the table above, we see that the point of intersection, to within a tolerance of 10^{-9}, is x=0.314923059.

Check it out

- Change the initial guess to $x_0=1$. How many iterations are needed before reaching the stopping criteria?

Suggested Exercises

Text	Section(s)	Exercises
Larson-Edwards *Calculus: An Applied Approach,* 6th ed.	10.6	#13-16
Berresford-Rockett *Applied Calculus,* 3rd ed.	10.4	#21-24,33-38